PHYSICS IN FOCUS

SKILLS AND ASSESSMENT
WORKBOOK

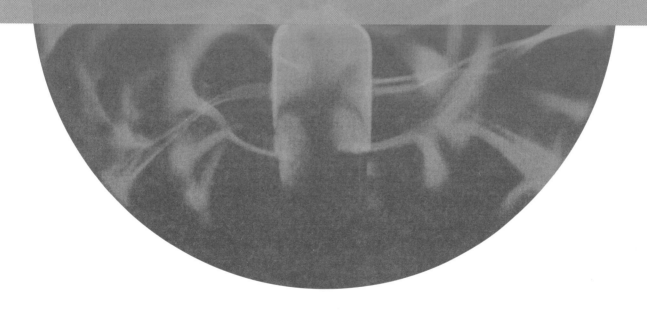

YEAR 11

Edward Baker
Darren Goossens
Owen Hamerton

Physics in Focus: Skills and Assessment Workbook Year 11
1st Edition
Edward Baker
Darren Goossens
Owen Hamerton

Publisher: Sam Bonwick
Project editor: Kathryn Coulehan
Editor: Elaine Cochrane
Proofreader: Marta Veroni
Cover design: Original cover design by Chris Starr (MakeWork), Adapted by
    Justin Lim
Text design: Ruth Comey (Flint Design)
Project designer: Justin Lim
Permissions researcher: Corrina Gilbert
Production controller: Alice Kane
Typeset by: MPS Limited

Any URLs contained in this publication were checked for currency during the
production process. Note, however, that the publisher cannot vouch for the
ongoing currency of URLs.

Acknowledgements
Cover image: iStock.com/Brostock
Inquiry questions on pages 22, 34, 57, 68, 83, 106, 119, 128, 142, 153, 168,
177 and 189 are from the Physics Stage 6 Syllabus © NSW Education
Standards Authority for and on behalf of the Crown in right of the State of
New South Wales, 2017.

For product information and technology assistance,
    in Australia call **1300 790 853**;
    in New Zealand call **0800 449 725**

For permission to use material from this text or product, please email
aust.permissions@cengage.com

ISBN 978 0 17 044959 5

Cengage Learning Australia
Level 7, 80 Dorcas Street
South Melbourne, Victoria Australia 3205

Cengage Learning New Zealand
Unit 4B Rosedale Office Park
331 Rosedale Road, Albany, North Shore 0632, NZ

For learning solutions, visit **cengage.com.au**

Printed in China by 1010 Printing International Limited.
1 2 3 4 5 6 7 24 23 22 21 20

CONTENTS

1 Working scientifically and depth studies

MODULE ONE » KINEMATICS

MODULE TWO » DYNAMICS

ABOUT THIS BOOK

FEATURES

▶ Introductory worksheets in Year 11 provide opportunities for you to learn good practices in working scientifically and prepare you to complete high-quality depth studies.

▶ Review prior knowledge from Stage 5 at the start of each module and check your understanding of key concepts at the end of each module.

▶ Learning goals are stated at the top of each worksheet to set the intention and help you understand what's required.

▶ Chapters clearly follow the sequence of the syllabus and are organised by inquiry question.

▶ Page references to the content-rich student books provide an integrated learning experience.

▶ Brief content summaries provided where applicable.

▶ Hint boxes provide guidance on how to answer questions effectively.

▶ Fully worked solutions appear at the back of each book to allow you to work independently and check your progress.

ORGANISATION OF YOUR WORKBOOK

An introduction to working scientifically and depth studies

This chapter focuses on the working scientifically skills. These skills are essential for success in the course and we walk you through each one to develop your understanding before applying them to the module content. You may also refer back to this chapter throughout the course.

Subsequent chapters

Each chapter begins with the relevant inquiry question and follows the sequence of the syllabus. Worksheets have been designed to complement the student book and provide additional opportunities to apply and revise your learning. Completion of these worksheets will provide you with a solid foundation to complete assessments and depth studies.

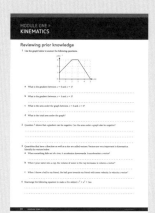

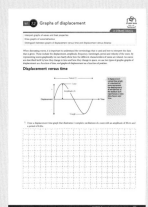

1 Working scientifically and depth studies

 Proposing a research question or hypothesis

STUDENT BOOK
Pages 11–12

> **LEARNING GOALS**
>
> Identify appropriate research questions and hypotheses
>
> Describe dependent and independent variables
>
> Describe the difference between a research question and a hypothesis

1 Choose the better research question and explain your choice.

   A What should car manufacturers do to make cars safer?

   B What features could be used in cars to make drivers safer in head-on collisions?

   _____

   _____

2 Think carefully about what a hypothesis is and what makes a good hypothesis. Write your explanation in the space below. Discuss and compare answers with a classmate.

   _____

   _____

   _____

   _____

   _____

3 Define 'independent variable'.

   _____

   _____

4 Define 'dependent variable'.

   _____

   _____

5 Explain why the following statement is an example of a good hypothesis.

   If the gradient of a slope is increased, then the velocity of a ball rolling down it will increase.

   _____

   _____

   _____

6 Explain why the following statement is an example of a poor hypothesis.

Is a lithium ion battery better than a lead acid battery?

_____

_____

7 If a student wanted to find out the best material to use to make bouncy balls, what would be a suitable hypothesis?

_____

8 What should you do if your investigation does not support your hypothesis?

_____

_____

_____

9 Owen and Doerte were looking at testing the differences in gaze behaviour between old and young people when they are walking up and down stairs. They hypothesised that older adults would look further ahead on stairs than younger people. When they tested several different subjects in different age groups to see if their hypothesis was supported, they found that there wasn't a statistical difference in where the different groups looked on the stairs. However, they did find that older people fixated on the steps for longer than younger people. Does this mean that their experiment did not work?

_____

_____

_____

_____

_____

10 What is the difference between a hypothesis and a research question?

_____

_____

_____

1.2 **Assessing risk in primary investigations**

STUDENT BOOK
Pages 14–15

Explain the difference between primary and secondary investigations

Identify risks in primary investigations

1 Two pairs of students each conduct an investigation. Identify which is the primary investigation and which is the secondary investigation. Explain.

A Jude and Natalie are carrying out an investigation into the effect of different insulators and how they can prevent heat loss to the environment. They research, set up and carry out an investigation, from which they record their data. They are able to come up with an answer based on their results.

B Hadja and Lachlan are carrying out an investigation into the development of the photoelectric effect and how it has helped with understanding in modern physics. They have undertaken extensive research into the work of Einstein and Planck and how it helped in understanding the processes involved. Hadja and Lachlan write up and present their work.

2 In Senior Physics, risk assessments must be specific to the investigation to be conducted.

Jason and Jo are testing the resistance in a parallel circuit compared to a series circuit. They are using a 12 V power supply, several connecting wires, three bulbs, two ammeters and three voltmeters.

Risks for their investigation are stated in the table below. Complete the risk assessment for the investigation.

What are the risks in doing this investigation?	How can you manage these risks to stay safe?
Short circuit in the power supply	
Light bulb may break	
Trip hazard over the wires	

3 Identify three risks for an investigation that involves moving a piece of wood up an inclined plane using weights and a pulley, and explain how the risks can be managed.

What are the risks in doing this investigation?	How can you manage these risks to stay safe?

LEARNING GOALS

Consider pros and cons when selecting equipment for investigations

Explain what makes investigations valid and reliable

Selecting equipment

When selecting equipment for a primary investigation, it is good practice to carry out a simple pros versus cons analysis. For example, the table below lists some pros and cons for using a stopwatch to measure time.

Pro	Con
Easily available	Relies on human reactions, which can be a major source of error over short time intervals
Easy to use	Difficult to measure start and stop times precisely

1 List the pros and cons for the technology below.

   a Data-logger

Pro	Con

   b Smart phone

Pro	Con

   c Metre ruler

Pro	Con

d Vernier calipers

Pro	Con

e Think of another technology that can be used to measure time or distance. It can be any type of technology as long as it is sensible and justified.

_____

Pro	Con

Validity and reliability

A group of students were seeking to test if a launch angle of 45° would give the greatest range for a projectile. They set up an experiment in which they changed the launch speed and the launch angle of the projectile and measured its range.

2 Was this a valid investigation? Explain.

_____

_____

_____

3 How could the investigation be changed to improve it?

_____

_____

4 How would the students ensure their experiment was reliable?

_____

_____

_____

WS 1.4 Selecting and referencing sources

Identify and evaluate reputable and non-reputable sources

Identify types of plagiarism and describe methods of avoiding them

Apply the rules of a referencing style

When carrying out a secondary-sourced investigation, it is important to use reputable sources for your information.

1 Consider the advantages of using a .gov.au website over a .com.au site and complete the table below.

.com.au		.gov.au	
Advantages	Disadvantages	Advantages	Disadvantages

2 Identify reputable sources below and state why they are reputable.

Type of source	Why is it reputable?

3 Identify some resources below that may not be reputable and should be considered carefully before use.

Type of source	Why is it not reputable?

In order to prevent accidental or deliberate plagiarism, the work of others must be referenced. There are several different referencing styles and each school will have a chosen referencing style. For this worksheet, we will use the (name, date) style of referencing, for example (Baker et al., 2020). For further information on this style of referencing, see APA 6th edition.

4 What are two types of plagiarism and how would you avoid them?

_____

_____

_____

_____

5 Re-write the original text to include correct in-text referencing.

Student work	Original text (Bloggs and Bright, 2019)
	The investigators found that when they dropped the ball from various heights the rebound of the ball changed due to the efficiency of the material used to make the ball. They were able to identify that vulcanised rubber was the best material to use for bouncy balls.

6 Maro was researching a topic for her depth study on waves. She found an article that she really liked online and wanted to use the information within it. She had the following information: the title, author, journal title, journal number, publication date and the page numbers. How would this be presented in the reference list?

_____

WS 1.5 Tabulating and graphing data

STUDENT BOOK
Pages 21–2

Explain the difference between quantitative and qualitative data

Process data into tabular and graphical forms

Tabulating data

1 Mariam was carrying out an investigation to see how the incline of a slope affected the speed of a ball coming down it. She had the following table of results.

Gradient (°)	Velocity (m s^{-1})
10	3
15	5
20	9
25	13

What type of data did Mariam have?

_____

2 Ross was carrying out an investigation to look at the reaction when he put sodium into water containing an indicator. Ross saw that the water turned purple and deduced that sodium hydroxide was being produced. What type of data did Ross use? Explain.

_____

_____

3 Within Physics, most of the quantitative data that you will be recording should be presented in either table or graphical form. A common pitfall for Physics students is not making sure that they have drawn tables and graphs correctly.

a Identify the problems with the table of results below.

Sports equipment	time		speed	
Cricket bat		15	23	
Tennis racket	34		17	
Hockey stick	20		19	
Baseball bat	11			28

_____

_____

b Place the following data of the displacement of a ball rolling down a slope into a table:

Time in seconds 1.0, 2.0, 3.0, 4.0, 5.0, 6.0, 7.0, 8.0, 9.0, 10

Displacement in cm 12, 14, 18, 26, 39, 50, 65, 82, 103, 115

Drawing graphs

4 Put the data from the table you created in part **b** of in question **3** into a graph.

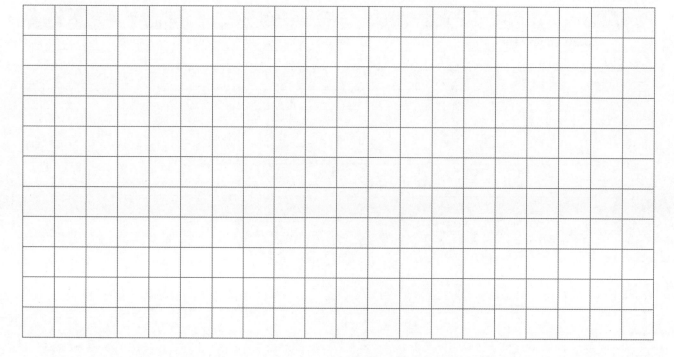

Measuring the gradient

If you have a linear graph (the data gives a straight line) then it may be necessary to find the gradient of the line. This tells you about the relationship between the dependent and independent variables.

To linearise data we must manipulate one of the variables. The graphs below are some of the graphs encountered in the Stage 6 Physics course. Below each graph is the equation that is required to linearise the data.

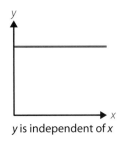

y is independent of x

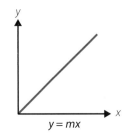

$y = mx$

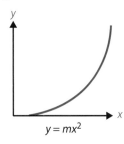

$y = mx^2$

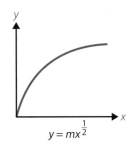

$y = mx^{\frac{1}{2}}$

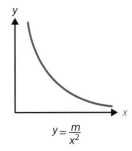

$y = \dfrac{m}{x^2}$

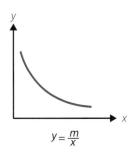

$y = \dfrac{m}{x}$

5 Sketch the shape of the following graphs.

  a $V \propto S$
  b $a \propto t^2$
  c V is independent of t

6 a Linearise the data tabulated in question **3b** and use it to plot a straight-line graph. In this case, the position depends upon the square of the time: $s \propto t^2$.

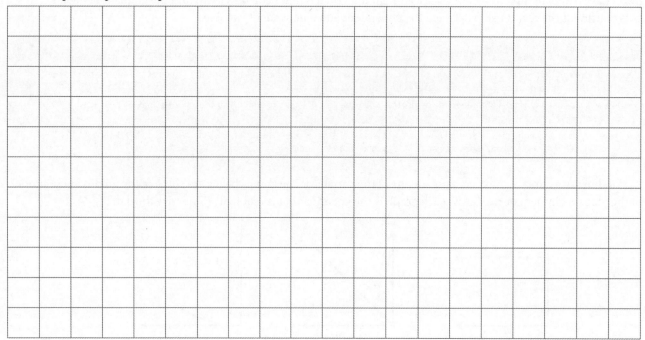

b Find the gradient of this graph.

_____

1.6 Identifying errors and uncertainties

 STUDENT BOOK
Pages 17–20

LEARNING GOALS

Identify and mitigate errors

Calculate uncertainty in measurements

Explain the importance of precision and accuracy in experiments

Error

Every measurement has an uncertainty, or error. Note that in this context 'error' does not mean mistake; it is the precision of the measurement. The less precise the measurements, the greater the error. Errors fall into two categories: systematic error and random error.

1 In the space below, describe what the error is and a way it can be mitigated.

   a Random error

   b Systematic error

2 How can a measuring device such as an electronic balance give both systematic and random errors?

Uncertainty

When reading an analogue scale, the uncertainty is half of the smallest division. So on a 30 cm ruler that has millimetres as its smallest division, the uncertainty is half millimetre or ±0.05 cm. It is not possible to know how digital devices round the measured value to get the display value so, for digital equipment, the uncertainty is the smallest scale division. These uncertainties are called 'limit of reading' uncertainties or resolution. It should not be assumed this is the only uncertainty due to a measuring device. Always read the manual and see if the uncertainty is larger than the resolution – often it is.

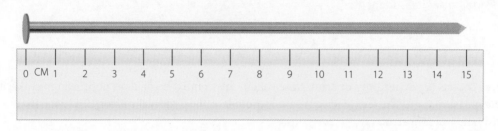

3 How long is the nail (including the uncertainty)?

_____

4 Within what range is the length of the nail with the above uncertainty?

_____

5 What would be the fractional uncertainty? Show the working.

_____

6 If you were to express the fractional uncertainty as a percentage uncertainty, what would it be?

_____

Combining uncertainties

It is necessary to combine the uncertainties in results correctly. Uncertainty can be shown in two ways: fractional and absolute.

When adding or subtracting measurements then you add the absolute error, for example if $R = a + b + c$ then the uncertainty in R would be $\Delta R = \Delta a + \Delta b + \Delta c$.

7 What would be the uncertainty in the following derived values for a rectangle measured to be 4.0 ± 0.2 m in length and 2.0 ± 0.3 m in width?

a Perimeter

_____

b Difference in the two side lengths

_____

When multiplying or dividing a measurement the fractional uncertainties are added. For example, if $R = \dfrac{ab}{c}$, the fractional uncertainty in R is $\dfrac{\Delta R}{R} = \dfrac{\Delta a}{a} + \dfrac{\Delta b}{b} + \dfrac{\Delta c}{c}$.

c What is the area of the rectangle?

_____

d What is the fractional uncertainty in the area?

_____

e What is the absolute uncertainty in the area?

_____

8 We calculate uncertainties to judge whether our experiment agrees with our hypothesis. If we hypothesise that the acceleration of a falling stone is $9.8\,\mathrm{m\,s^{-2}}$, which of the following results supports this hypothesis? Explain.

    A $(9.78 \pm 0.01)\,\mathrm{m\,s^{-2}}$

    B $(9.9 \pm 0.1)\,\mathrm{m\,s^{-2}}$

    C $(12 \pm 2)\,\mathrm{m\,s^{-2}}$

Precision and accuracy

9 A GPS system is trying to triangulate where Katy is to enable her to find out how to get to a concert venue. The GPS system works by using different recordings of a receiver in relation to several satellites. Explain the importance of precision and accuracy in this situation.

WS 1.7 Using modelling to solve problems

Understand how models change over time

Identify appropriate models for particular situations

An example of modelling in physics is the building of the atomic model and how this has helped our understanding of the atom. Each stage of the model built on the stage before it.

1 What is one of the major driving forces for improvements on models?

_____

_____

2 An early change to the atomic model was made by Ernest Rutherford. On what evidence did Rutherford base his change to the model, and what was the change?

_____

_____

_____

_____

3 The next change to the model was when Neils Bohr identified that electrons travelled around the nucleus in discrete energy levels. What caused Bohr to investigate this further?

_____

_____

_____

_____

4 In the following situations, which type of model - mathematical or physical – would be more suitable, and why?

a Finding out what happened at the time of the Big Bang

_____

_____

b Modelling the expansion of the universe

_____

_____

c To see what would happen in a car crash

_____

_____

WS 1.8 Presenting information

LEARNING GOALS

Consider the pros and cons of ways to present information

Describe the sections of a formal report

1 Construct a table that shows the pros and cons of the following ways to present your information: video, PowerPoint or oral presentation, formal written report.

2 A formal report has several sections, each with a different function.

- Abstract
- Introduction
- Method
- Results and analysis
- Discussion
- Conclusion
- Acknowledgments
- Reference list

a What is the purpose of the abstract?

_____

_____

_____

b What is the purpose of the introduction?

_____

_____

_____

c Which method would be the correct one from the two below?

    A   Gather equipment.
            Fill up a beaker with water.
            Turn on Bunsen burner.
            Put the beaker on the Bunsen burner.
            Put thermometer in the beaker.
            Measure the temperature change.

    B   The beaker was filled with 100 mL of water.
            The beaker was placed on the tripod over a lit Bunsen burner.
            The temperature change was measured at 30 second intervals.
            The results were recorded in a table.

In the space below, make your choice and justify your answer.

_____

_____

_____

d If you have achieved results in your investigation, it is important to display them. How should you display your results?

_____

_____

_____

e Why is it important to analyse your results?

_____

_____

_____

f Which of these discussion points would be appropriate to include in this section? Justify your answer.

    A   When measuring the speed of the cart down the incline, we struggled to get accurate timing data. We could have improved this by making the incline longer to reduce the effects of reaction rates on the time.

    B   Our results were not as accurate as they could have been due to the timing being controlled by humans. The best way to deal with this would have been for the school to provide automated timing equipment.

_____

_____

_____

g Choose which conclusion is in the more suitable format and state why.

    A   Looking at our results we showed that they support our hypothesis as when we increased A then B happened. This shows that it is a valid statement.

    B   Our results were good and they showed what we expected to happen. The experiment was carried out well and nothing went wrong.

_____

_____

_____

h Who might you thank in your acknowledgments section?

_____

i What is the difference between a reference list and a bibliography?

_____

_____

_____

Reviewing prior knowledge

1 Use the graph below to answer the following questions.

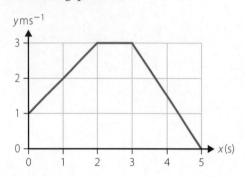

   a What is the gradient between $x = 0$ and $x = 2$?

   b What is the gradient between $x = 3$ and $x = 5$?

   c What is the area under the graph between $x = 0$ and $x = 2$?

   d What is the total area under the graph?

2 Question **1** shows that a gradient can be negative. Can the area under a graph also be negative?

3 Quantities that have a direction as well as a size are called vectors. Vectors are very important in kinematics. Identify the vectors below.

   a When something falls out of a tree, it accelerates downwards. Is acceleration a vector?

   b When I pour water into a cup, the volume of water in the cup increases. Is volume a vector?

   c When I throw a ball to my friend, the ball goes towards my friend with some velocity. Is velocity a vector?

4 Rearrange the following equation to make a the subject: $v^2 = u^2 + 2as$.

5 Use trigonometry to find the unknown values in this right-angled triangle.

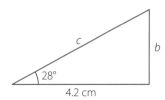

a What is the length of *c*?

_____

b What is the length of *b*?

_____

c Use Pythagoras's theorem to check your results.

_____

_____

_____

2 Motion in a straight line

INQUIRY QUESTION: HOW IS THE MOTION OF AN OBJECT MOVING IN A STRAIGHT LINE DESCRIBED AND PREDICTED?

 Describing motion

STUDENT BOOK
Pages 31–7

LEARNING GOALS

Describe qualitatively uniform motion and uniformly accelerated motion effectively using scientific terminology

We move about all the time, by walking and in cars, trains and aeroplanes. Even when you are sitting still, much of your body is moving; blood is flowing through your veins and arteries, and inside every cell proteins and sugars are constantly on the move.

To describe motion, we use the terms time, distance, displacement, speed, velocity and acceleration. Some of these terms refer to scalar quantities and some to vector quantities.

1 What is the difference between a scalar and a vector? Give an example of each.

_____

_____

_____

_____

2 Complete the table below to show which of these quantities are vectors and which are scalars, and give the SI units for each.

Quantity	Scalar or vector?	SI unit
time		
distance		
displacement		
speed		
velocity		
acceleration		

3 Sam travels a total distance of 36 km on Monday. What is his maximum possible displacement in that day, and what is his minimum possible displacement? Explain how each of these is possible.

_____

_____

_____

_____

_____

4 Explain the difference between speed and velocity.

_____

_____

_____

_____

_____

5 An object is accelerating in a straight line. Is its velocity necessarily in the same direction as its acceleration? If not, give an example of when velocity and acceleration are in opposite directions.

_____

_____

_____

_____

6 Write an equation that relates velocity to displacement and time. Explain in words what this equation means.

_____

_____

_____

_____

Investigate average velocity and instantaneous velocity through a practical with an oscillating system

INVESTIGATION

Average velocity and instantaneous velocity for an oscillating object

AIM

To investigate the average and instantaneous velocity of an object oscillating on the end of a spring

HYPOTHESIS

1 Write a hypothesis for this experiment. Think about how you expect the instantaneous velocity to vary.

_____

_____

_____

RISK ASSESSMENT

2 Complete a risk assessment that covers at least two risks that are specifically associated with this investigation.

What are the risks in doing this investigation?	How can you manage these risks to stay safe?

MATERIALS

• Spring   • Small object to hang on spring   • Ruler   • Stopwatch and video camera *or* 2 smart phones

METHOD

1 Position the ruler vertically against a wall.

2 Put the stopwatch (or smart phone displaying stopwatch) next to the ruler.

3 Hang the small object on the spring and hold it in front of the ruler.

4 Position the video camera (or smart phone in video mode) in front of the spring, ruler and stopwatch so that it has a clear view of the object, ruler and stopwatch. You need to be able to read the stopwatch and the position of the object in the picture.

5 Start the stopwatch and video camera.

6 Pull down on the object and record its motion for at least three oscillations. An oscillation is a complete cycle of the motion, for example, from the top to the bottom and back again.

> **HINT**
>
> If the oscillations are too fast for the frame rate of the camera to give good data, replace your oscillating object with a heavier one.

3 View the recording of the motion, stepping through as slowly as possible. Draw a graph of the position of the object as a function of time on the axes below. Choose a sensible scale so that your graph fills the space as much as possible.

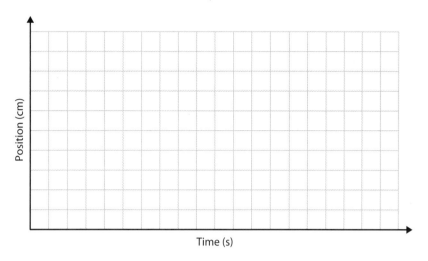

ANALYSIS

4 Mark on your graph where the magnitude of the instantaneous velocity is maximum and where it is minimum.

5 What is the value of the minimum instantaneous velocity?

_____

6 The instantaneous velocity is calculated by finding the gradient of the tangent to the graph at the instant of interest. Draw a tangent to the graph at the point of maximum velocity. Use this line to calculate the maximum instantaneous velocity. Remember to keep units on all values as you work!

_____

_____

7 Choose a time period for which the average velocity is not zero. Mark this time period on your graph and calculate the average velocity for this time period.

_____

_____

8 Over what time period/s is the average velocity zero? Does it matter when this time period starts?

_____

_____

_____

9   Consider the following questions and discuss them with your group. Use the space below to make notes.

   a   How does the instantaneous velocity vary with time for this oscillating object?

   b   How does the average velocity depend on the time period that you choose?

   c   If you were going to carry out this investigation again, what changes would you make to ensure a more valid/reliable investigation?

   _____

   _____

   _____

   _____

   _____

   _____

   _____

   _____

   _____

   _____

   _____

   _____

CONCLUSION

10 Did your results support your hypothesis? Write a suitable conclusion to your investigation, making sure that you relate it back to your hypothesis.

   _____

   _____

   _____

   _____

   _____

   _____

   _____

   _____

   _____

   _____

   _____

   _____

WS 2.3 Calculating relative velocity

Calculate relative velocity in a variety of one-dimensional situations

If you are running at $10\,\mathrm{m\,s^{-1}}$, and someone is running towards you at $10\,\mathrm{m\,s^{-1}}$, then they are getting closer to you at a rate of 20 m every second. Both of you are moving at more than $1000\,\mathrm{km\,h^{-1}}$ around the centre of Earth, and at more $100\,000\,\mathrm{km\,h^{-1}}$ around the Sun. So how fast are you moving? Well, it depends who you ask, and what their reference frame is.

1 Define 'reference frame'.

_____

_____

_____

_____

2 What reference frame does the speedometer of a car measure the speed of the car relative to?

_____

3 The velocity of object A relative to object B is given by $\vec{v}_A - \vec{v}_B$, where both these velocities are measured relative to some reference frame or point. This is the velocity that object B sees object A travelling at. Complete the table below to show when the relative velocity of A to B is positive, negative or zero, and when the objects are getting closer together or further apart.

Velocities of A and B	$\vec{v}_A - \vec{v}_B$ positive or negative?	A and B getting closer together or further apart?
$\vec{v}_A$ positive and $\vec{v}_B$ negative, position of A is positive and position of B is negative		
$\vec{v}_A$ negative and $\vec{v}_B$ positive, position of A is positive and position of B is negative		
$\vec{v}_A$ and $\vec{v}_B$ both positive and equal		
$\vec{v}_A$ and $\vec{v}_B$ both positive, $\vec{v}_A$ greater than $\vec{v}_B$ and A has the larger displacement from the reference point		
$\vec{v}_A$ and $\vec{v}_B$ both positive, $\vec{v}_A$ greater than $\vec{v}_B$ and B has the larger displacement from the reference point		

4 In the figures below, the velocities of A and B are represented by vectors with length proportional to the speed of the object. For each of the figures below, state whether the relative velocities $\vec{v}_A - \vec{v}_B$ and $\vec{v}_B - \vec{v}_A$ are positive or negative.

a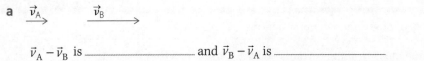

$\vec{v}_A - \vec{v}_B$ is _____ and $\vec{v}_B - \vec{v}_A$ is _____

b $\vec{v}_A$ _____→ $\vec{v}_B$ __→

$\vec{v}_A - \vec{v}_B$ is _____ and $\vec{v}_B - \vec{v}_A$ is _____

c ←_ $\vec{v}_A$ ←_ $\vec{v}_B$ ____

$\vec{v}_A - \vec{v}_B$ is _____ and $\vec{v}_B - \vec{v}_A$ is _____

d $\vec{v}_A$ __→ $\vec{v}_B$ ←__

$\vec{v}_A - \vec{v}_B$ is _____ and $\vec{v}_B - \vec{v}_A$ is _____

5 When you are travelling in a car, are the relative speeds (to you) of other cars going in the same direction greater or smaller than the speeds shown on their speedometers? Why? What about for cars going in the opposite direction?

_____

_____

_____

_____

_____

_____

6 Olga is riding her bike at $40\,\text{km}\,\text{h}^{-1}$ along a bike lane when she is overtaken by a car travelling at $60\,\text{km}\,\text{h}^{-1}$ in the car lane next to her. Assume they are both travelling in the positive direction.

a What is the speed of the car relative to Olga as it approaches her?

_____

b What is the speed of Olga relative to the car as it approaches her?

_____

c Would your answers be any different for the relative speeds as the car pulls ahead of Olga?

_____

Describing motion with the kinematics equations

LEARNING GOALS

Analyse relationships between time, distance, speed, velocity and acceleration through equations and graphs

To describe motion, we use the quantities time, distance, displacement, speed, velocity and acceleration. These quantities are related mathematically. Speed is the rate of change of distance and velocity is the rate of change of displacement. Speed is the magnitude of the velocity vector. Acceleration is the rate of change of velocity.

We write these relationships mathematically as:

$$v = \frac{s}{\Delta t}, \quad \vec{v} = \frac{\vec{s}}{\Delta t} \text{ and } \vec{a} = \frac{\Delta \vec{v}}{\Delta t}$$

where $\vec{v}$ is the velocity vector, and v is its magnitude, the speed. The vector $\vec{s}$ is the displacement and s or d is its magnitude, the distance. The vector $\vec{a}$ is the acceleration. The symbol Δ means a change in a quantity, so Δt means the change in time, or the time interval.

1 Lei walks 200 m in 5 minutes. What is his speed? Give your answer in $m\,s^{-1}$.

_____

_____

2 Lei starts at rest and accelerates uniformly until he is running at $8.6\,m\,s^{-1}$. It takes him 2 s to reach this speed. Calculate his acceleration.

_____

_____

_____

3 It takes Lei 28 s to run around a circular track with circumference 200 m.

a What distance has Lei travelled?

_____

b What is Lei's speed as he runs?

_____

c What is Lei's displacement?

_____

d What is Lei's average velocity over the lap?

_____

4 Elena walks at a speed of $1.4\,m\,s^{-1}$.

a How long will it take her to travel 500 m?

_____

_____

b How far will she travel if she walks for 2 hours? Give your answer in units of km.

_____

_____

_____

5 If a car can brake with a uniform acceleration of $11\,\mathrm{m\,s^{-2}}$, how long will it take to come to a stop from $100\,\mathrm{km\,h^{-1}}$?

_____

_____

_____

_____

6 Every morning Ahmed walks briskly to the nearest bus stop and waits patiently for the bus. He then takes the bus to the closest stop to his school and walks the rest of the way. This motion is shown on the graph below.

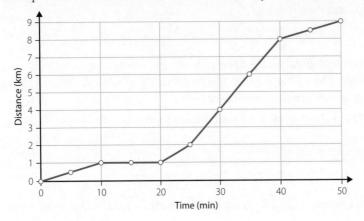

a What is Ahmed's total displacement during this trip?

_____

b What is Ahmed's average speed during this trip?

_____

_____

c Mark on the graph where Ahmed's maximum and minimum speeds occur.

d Calculate Ahmed's maximum speed.

_____

_____

_____

7 On the axes below, plot a graph of Ahmed's speed as a function of time during his trip to school. Make sure you convert units correctly!

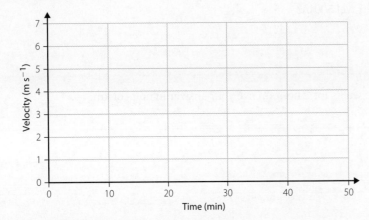

8 The graph below shows the velocity as a function of time for an aeroplane flying from Sydney to Melbourne.

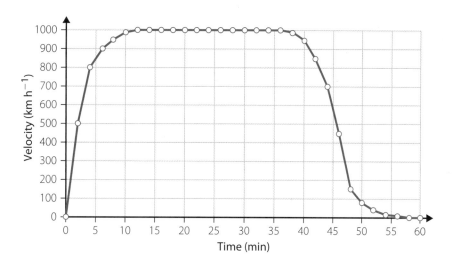

a Calculate the maximum acceleration of this aeroplane. Give your answer in $m\,s^{-2}$.

b Estimate, using the graph, the total displacement of the aeroplane during this flight.

> **HINT**
>
> The area under a curve can be quickly estimated by counting the rectangles under the curve. If there are roughly equal numbers of partial rectangles larger than and smaller than half a full rectangle, then you can just count the number of partial rectangles and divide by two. This gives you the approximate number of rectangles contributed by the partial rectangles.

c Using this total displacement, calculate the average velocity of the aeroplane during this flight. Give your answer in $km\,h^{-1}$ and $m\,s^{-1}$.

More equations for describing rectilinear motion

STUDENT BOOK
Pages 46–50

Derive and use the equations of motion for uniformly accelerated objects

Starting with the mathematical relationships:

$$v = \frac{s}{\Delta t}, \quad \vec{v} = \frac{\vec{s}}{\Delta t} \quad \text{and} \quad \vec{a} = \frac{\Delta \vec{v}}{\Delta t}$$

we can derive some more useful equations for describing motion, when we assume that the acceleration is constant:

$$\vec{s} = \vec{u}t + \frac{1}{2}\vec{a}t^2, \quad \vec{v} = \vec{u} + \vec{a}t \quad \text{and} \quad \vec{v}^2 = \vec{u}^2 + 2\vec{a}\vec{s}$$

where $\vec{u}$ is the initial velocity, so $\Delta \vec{v} = \vec{v} - \vec{u}$.

1 Show how $\vec{v} = \vec{u} + \vec{a}t$ can be derived from the equation $\vec{a} = \frac{\Delta \vec{v}}{\Delta t}$.

_____

_____

2 Follow these steps to show how $\vec{v}^2 = \vec{u}^2 + 2\vec{a}\vec{s}$ can be derived from the equations $\vec{a} = \frac{\Delta \vec{v}}{\Delta t}$ and $\vec{v} = \frac{\vec{s}}{\Delta t}$.

a Rearrange $\vec{a} = \frac{\Delta \vec{v}}{\Delta t}$ for Δt:

_____

_____

b Replace $\Delta \vec{v}$ with $\vec{v} - \vec{u}$ in the equation from part **a**.

_____

_____

When the acceleration is constant, the average velocity is given by $\vec{v}_{\text{ave}} = \frac{\vec{v} + \vec{u}}{2}$, so the displacement in any time Δt is $\vec{s} = \vec{v}_{\text{ave}}\Delta t = \frac{\vec{v} + \vec{u}}{2}\Delta t$.

c Substitute your expression for Δt from part **b** into $\vec{s} = \frac{\vec{v} + \vec{u}}{2}\Delta t$.

_____

_____

d Noting that for any a and b, $(a+b)(a-b) = a^2 - b^2$, rearrange your equation from part **c** to give $\vec{v}^2 = \vec{u}^2 + 2\vec{a}\vec{s}$.

_____

_____

_____

3 The acceleration due to gravity near Earth's surface is $9.8\,\text{m s}^{-2}$.

a If a ball is dropped from a height of 2.0 m, how long will it take to fall to the ground?

_____

_____

b On the Moon, the acceleration due to gravity is only $1.6\,\mathrm{m\,s^{-2}}$. How long would it take a ball dropped from 2.0 m above the Moon's surface to hit the ground?

_____

_____

c How high above Earth's surface would an object have to be for it to take this long to fall to the ground?

_____

_____

4 Eleanor is in the entrance lane for the Hume Highway at Yass, travelling at $60\,\mathrm{km\,h^{-1}}$. In 200 m she wants to be travelling at $100\,\mathrm{km\,h^{-1}}$ to merge safely onto the highway. Calculate her required acceleration.

_____

_____

_____

5 If Eleanor continues to accelerate at this rate after reaching $100\,\mathrm{km\,h^{-1}}$, how long will it take her to reach the speed limit of $110\,\mathrm{km\,h^{-1}}$?

_____

_____

_____

6 It takes Eleanor 6 s to accelerate from $100\,\mathrm{km\,h^{-1}}$ to $60\,\mathrm{km\,h^{-1}}$ when she leaves the Hume Highway at Goulburn. Assuming uniform acceleration, what distance does she travel in this time?

_____

_____

_____

_____

_____

_____

INQUIRY QUESTION: HOW IS THE MOTION OF AN OBJECT THAT CHANGES ITS DIRECTION OF MOVEMENT ON A PLANE DESCRIBED?

 3.1 ## Working with vectors in one and two dimensions

STUDENT BOOK
Pages 58–63

LEARNING GOALS

Gain familiarity with the use of vectors to represent quantities with magnitude and direction

Gain an understanding of how the components of a vector combine to give the vector

Be able to use a variety of methods to calculate some aspects of a vector (e.g. components, angles) from given information

Most of our motion takes place on a plane. While Earth is actually spherical and its surface is not smooth, over short distances we can approximate the ground as a plane.

Many of the quantities we use to describe and measure motion are vectors. In one dimension we can specify direction with just the sign of a vector, but in two dimensions we need to give more information to define direction.

1 Is it possible for the magnitude of a vector to be smaller than that of either of its components? Explain your answer.

_____

_____

_____

2 Is it possible for the magnitude of a vector to be equal to that of either of its components? Explain your answer.

_____

_____

_____

3 A vector $\vec{a}$ has length 35 cm. The magnitude of its x component $a_x = |\vec{a}_x| = 28$ cm.

  a Calculate the magnitude of the y component, $a_y = |\vec{a}_y|$.

_____

_____

_____

  b Draw the vector $\vec{a}$ and its components.

c Calculate the angles between $\vec{a}$ and $\vec{a}_x$ and between $\vec{a}$ and $\vec{a}_y$. Mark them on your diagram above.

_____

_____

4 A vector $\vec{s}$ has length 15 cm. Take the x direction as horizontal and the y direction as vertical. If $\vec{s}$ is at an angle of 45° to $\vec{s}_x$, calculate the values of $\vec{s}_x$ and $\vec{s}_y$.

_____

_____

_____

5 A vector $\vec{d}$ has length 200 km and is at an angle $\theta = 30°$ south of west.

a Draw this vector on the axes below.

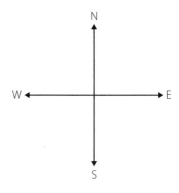

b Mark the components of $\vec{d}$ in the south and west directions on the diagram.

c Calculate the magnitudes of these components and mark them on the diagram.

_____

_____

_____

6 A vector has components $\vec{s}_x = 15$ m and $\vec{s}_y = 20$ m.

a Draw the vector $\vec{s}$ and its components.

b Calculate the length of the vector $\vec{s} = \vec{s}_x + \vec{s}_y$.

_____

_____

c Assuming the x direction is horizontal and the y direction is vertical, calculate the angle $\vec{s}$ makes to the horizontal.

_____

_____

_____

7 A vector $\vec{b}$ has x component $\vec{b}_x = 150\,\text{km}$, and the angle, θ, between $\vec{b}$ and $\vec{b}_x$ is $\theta = 30°$. Calculate the lengths of $\vec{b}$ and $\vec{b}_y$.

_____

_____

_____

_____

8 For each of the vectors below, calculate and write in the unknown components.

a
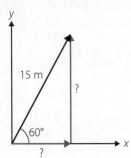

b
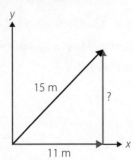

c
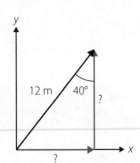

d
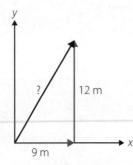

_____

_____

_____

_____

_____

_____

_____

WS 3.2 Distance and displacement in two dimensions

STUDENT BOOK
Pages 64–70

Displacement is how far and in what direction something has moved from its previous position. It is a vector. Let us say our initial (i) position relative to the origin was given by vector $\vec{d}_i$ and our final (f) position by $\vec{d}_f$. Our displacement in moving from $\vec{d}_i$ to $\vec{d}_f$ is given by the difference between these two positions. We use *s* for displacement, and it is a vector, so $\vec{s} = \vec{d}_f - \vec{d}_i$. Note that the path we took does not matter. If we drive north 100 km then south 100 km, our displacement *at the end* of the trip is zero, even though our path was 200 km long.

Similarly, we can add displacements to get a final displacement. This is vector addition. When adding mathematically, we resolve the vectors into perpendicular components, add the components and then combine the results to give the resultant vector. We can also add vectors graphically.

1 I begin a trip and I drive 100 km straight north from my starting point.

   a Draw a set of axes (N, S, E, W) with a consistent scale, and mark in my position as a vector.

   b Write $\vec{s}$ in vector notation.

   _____

   I continue my trip. I now drive 100 km straight south-west.

   c Draw this leg of the trip on your set of axes.

   d What *distance* have I driven in total?

   _____

   e By measuring off your diagram, find my total displacement – that is, the vector sum of the two legs of the trip. Mark the vector on your diagram as $\vec{R}$, the resultant, and write it here as a length and direction.

   _____

   f Which is longer, total displacement or total distance travelled? Do you think it could ever be the other way around? How could the magnitudes of the two quantities come out equal?

   _____

   _____

   _____

2 Graphically add these pairs of displacement vectors. Find $\vec{R}$ where $\vec{R} = \vec{A} + \vec{B}$. The first one has been done for you.

a

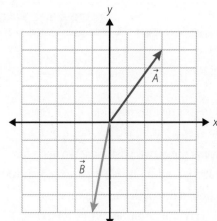

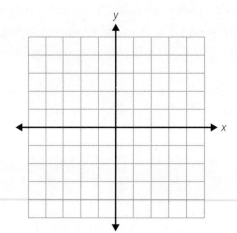

b

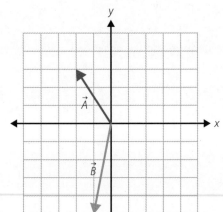

c

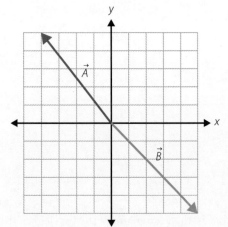

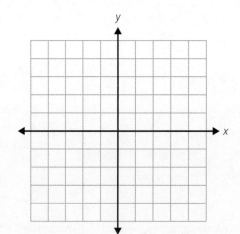

3 Graphically subtract these pairs of displacement vectors. That is, find $\vec{R} = \vec{A} - \vec{B}$.

a

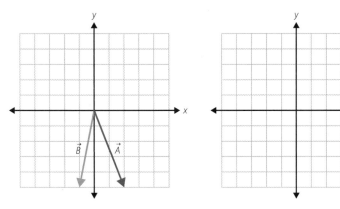

b

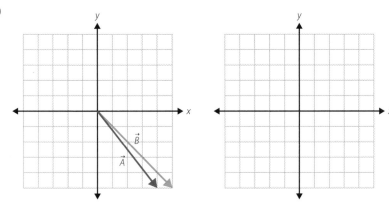

4 A cyclist rides 34 km north-east ($\vec{A}$) and then 20 km south ($\vec{B}$).

   a Add these two displacements to calculate their total displacement.

   _____

   _____

   _____

   _____

   _____

   _____

   b Write down a vector that describes their shortest possible path back to where they started.

   _____

5 While orienteering, Maya runs 1.00 km in a direction 27° east of south (S27.0°E) ($\vec{A}$). She decides she's done enough running, and so walks 800 m west ($\vec{B}$). What is her displacement relative to where she started the first leg? (A diagram is a good place to start.)

_____

_____

_____

_____

_____

6 Bryan is at a point 350 m north and 200 m east of the ice-cream shop. We could write his x and y coordinates as a Cartesian (x, y) pair: (350 m, 200 m). Julie is 200 m north and 350 m west of the shop – at coordinates (200 m, –350 m). What is Julie's displacement relative to Bryan?

a Write your answer as a pair of coordinates in brackets, like (x, y).

_____

_____

b Work out your answer as a vector with a length and a direction.

_____

_____

_____

7 Sandeep is 600 m from the ice-cream shop, in a direction N25°E.

a How far to the east of the shop is he?   **b** How far to the north is he?

_____   _____

_____   _____

_____   _____

_____   _____

Małgosia is 500 m from the shop in a direction S40°W.

c How far west of Sandeep is Małgosia?   **d** How far south of Sandeep is Małgosia?

_____   _____

_____   _____

_____   _____

e How far from Sandeep is Małgosia?

_____

_____

_____

_____

STUDENT BOOK
Pages 70–7

Model how motion can change in two dimensions, for example by comparing velocities at two different points in the motion

Apply constant acceleration equations in two dimensions

See what happens when we have motion in two dimensions and there is an acceleration present

The motion of an object changes if it accelerates. If an object has a velocity, its displacement is changing. If an object is accelerating, its velocity is changing.

Velocity is a vector. It gives the speed and direction of the motion. If we write the velocity as $\vec{v}$, we may write the speed as v. When we move from one point to another, we can find an average velocity for the journey. It is given by the total displacement divided by the total time. This may be different from the actual velocities experienced during the motion. We can combine component velocities to get a final velocity just as we combine components of displacement.

The rules for constant acceleration (chapter 2) apply in the plane.

1 Rashid drives 100 km straight north from his starting point (vector $\vec{A}$). He drives at 100 km h^{-1}. He then drives 120 km straight south-west at 80 km h^{-1} (vector $\vec{B}$).

   a What *distance* has he driven in total?

   b What was his average speed?

   c Find his total displacement and work out his average velocity for the whole trip.

   d Which is larger, average speed or average velocity? Do you think it could ever be the other way around? How could the magnitudes of the two quantities come out equal?

2 Allen rides 34 km north-east ($\vec{A}$) and then 20 km south in a total of 60 minutes ($\vec{B}$).

Note: You may have calculated the displacement in worksheet 3.2 question **4a**. You may use that answer to help you complete the following question.

a What velocity must Allen average over the next (and last) leg if he is to get back to where he started such that the total trip takes 80 minutes?

b What speed must he average over that last leg if he is to get back to where he started such that the average speed for the trip is 50 km h^{-1}? Recall that speed in this case is $\dfrac{\text{distance}}{\text{time}}$, not $\dfrac{\text{displacement}}{\text{time}}$.

3 While orienteering, I run at 12 km h^{-1} in a direction 27° east of south (S27°E) for 15 minutes ($\vec{A}$). I then walk at 6.0 km h^{-1} west for 45 minutes ($\vec{B}$).

a What was my average speed for the two legs combined?

b What was my average velocity for the two legs combined?

4 A ball travelling at a speed of 8.0 m s^{-1} hits a wall perpendicularly and bounces back along its own path. But it has lost some energy, so its speed is now 6.0 m s^{-1}.

a If we define 'towards the wall' as the positive direction, what is the change in velocity of the ball?

b What is the average acceleration the ball undergoes in the 2 s period beginning 1 s before it hits the wall?

c Will this average acceleration be more or less than the maximum acceleration the ball experiences during its bounce? Explain.

d Sketch a graph of the velocity as a function of time.

e Sketch a graph of acceleration as a function of time. Do not put a value on the vertical scale because we do not actually know how long the collision took.

5 Lauren has a yoyo. She is swinging it around her head in a horizontal circle. When it is in front of her, the yoyo is going to the left at $12\,\mathrm{m\,s^{-1}}$. When it is behind her, it is going to the right at $12\,\mathrm{m\,s^{-1}}$.

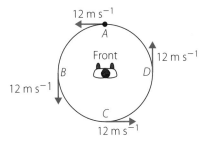

a What is the change in speed of the yoyo when it goes from in front (A) to behind (C)?

_____

b What is the change in velocity of the yoyo when it goes from A to C?

_____

_____

c What is the change in speed of the yoyo when it goes from A to B?

_____

d What is the change in velocity of the yoyo when it goes from A to B?

_____

_____

_____

_____

_____

_____

Extension

6 An air-hockey puck is sliding across a sloping, frictionless table. The puck's initial velocity, $\vec{v}_i$, has no component down the slope, but gravity will pull it down the slope with an acceleration of magnitude a.

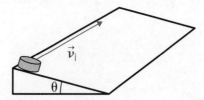

a Find an expression in terms of symbols t (time) and a (acceleration down the slope) for the magnitude of the component of the velocity down (parallel with, ∥) the slope. Call it $v_\parallel$.

_____

b What do we know about $v_\perp$, the velocity across (perpendicular to, ⊥) the slope?

_____

c Using this information, find an expression for the speed as a function of time.

_____

_____

_____

_____

d Find an expression for the direction of the velocity (that is, θ) as a function of time. Take the direction perpendicular to the slope – that is, the direction of $\vec{v}_i$ – as the zero of angle, so θ is the angle to that direction.

_____

_____

_____

e What is the value of θ at very small times? At very large times?

_____

_____

_____

_____

_____

Be able to work out the position of an object relative to that of another

Be able to model motion that has several components

Be able to combine velocities

Be able to work out the velocity of an object relative to that of another

When objects have positions given in some coordinate system, such as compass points or the Cartesian plane, they each have a displacement relative to the origin. They also have displacements relative to each other. Because displacement is a vector, we can get the displacement of one object relative to another by using vector analysis. If one object is at $\vec{d}_1$ and a second at $\vec{d}_2$, then the displacement of 1 relative to 2 is $\vec{s}_{1\ \text{relative to}\ 2} = \vec{d}_1 - \vec{d}_2$.

This is the same as saying that $\vec{s}_{1\ \text{relative to}\ 2}$ will take you from point 2 to point 1.

When two objects are in motion, both have a velocity relative to the origin of the coordinate system. They also have velocity relative to each other. Because velocity is a vector, we can get the velocity of one object relative to another by using vector subtraction:

$\vec{v}_{1\ \text{relative to}\ 2} = \vec{v}_1 - \vec{v}_2$.

1 Fill in the following table, qualitatively. There may be more than one answer. As an example, the first entry has been done for you.

Describe a situation in which …	Description
The displacement of one object relative to the other is constant	They have the same velocity (they are moving in parallel directions and keeping the same speed; $v = 0$ for both is one case).
The displacement of one object relative to the other is increasing at a constant rate (that is, linearly)	
The displacement of one object relative to the other is increasing at an increasing rate	
The objects have velocities opposite in direction and the relative displacement is decreasing	

2 My car (M) is parked in the fourth parking space from the wall. Each space is 2.5 m wide. Kaden's car (K) is parked in the tenth space from the wall. If distance increases as we move away from the wall, then:

   a   What is the position of Kaden's car relative to mine? (Give your answer in metres.)

   _____

   _____

   b   What is the position of my car relative to Kaden's? (Give your answer both in metres and in car spaces.)

   _____

3 Sunil drives 100 km straight north from his starting point. Mithali starts at the same point, and she also drives 100 km straight north. They meet at a café for lunch. Sunil then drives 100 km straight south-west. Mithali drives 80 km N30°E. What is Sunil's displacement relative to Mithali's?

   _____

   _____

   _____

   _____

   _____

   _____

4 Lauren has a yoyo. She is running east at $15\,\mathrm{km\,h^{-1}}$ and at the same time the yoyo moves up at $8.0\,\mathrm{m\,s^{-1}}$ relative to her hand. Calculate the velocity of the yoyo (Y) relative to the ground (G). Note: define the angle relative to the vertical.

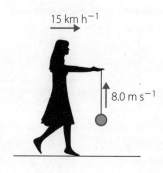

_____

_____

_____

_____

_____

_____

5 While continuing to run east at $15\,\mathrm{km\,h^{-1}}$, Lauren starts swinging the yoyo around her head in a horizontal circle. The yoyo has a tangential speed of $12\,\mathrm{m\,s^{-1}}$.

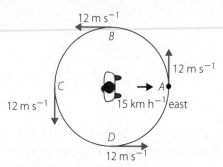

a Calculate the velocity of the yoyo relative to the ground when the yoyo is at A.

_____

_____

_____

_____

b Calculate the change in *speed* of the yoyo relative to the ground when it goes from in front (A) to behind (C).

_____

c Calculate the change in velocity of the yoyo relative to the ground when it goes from A to C.

_____

_____

d Calculate the change in speed of the yoyo relative to the ground when it goes from A to B.

_____

_____

e Calculate the change in velocity of the yoyo relative to the ground when it goes from *A* to *B*.

_____

_____

_____

_____

_____

6 Lauren is still running east at $15\,\text{km}\,\text{h}^{-1}$ and swinging the yoyo around her head in a horizontal circle. The yoyo has a tangential speed of $12\,\text{m}\,\text{s}^{-1}$. Consider the points *E* and *F*, at 45° to the E–W direction.

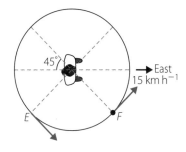

a What is the velocity of the yoyo relative to the ground at *E*?

_____

_____

_____

_____

_____

b What is the velocity of the yoyo relative to the ground at *F*?

_____

_____

_____

c What is the change in velocity of the yoyo when it goes from *E* to *F*?

_____

_____

_____

LEARNING GOALS

Be able to analyse situations in which objects are moving relative to each other in two dimensions

Be able to represent these situations using diagrams and equations

To explore motion relative to more than one frame of reference; for example, a boat relative to the water and relative to the bank of the river.

In previous worksheets we explored relative position and motion. In this section, we focus on relative motion, and explore several important examples – a boat on a river, two moving vehicles on the ground, and an aeroplane in a crosswind.

1 An aeroplane (A) is flying west. It is flying at $700\,\mathrm{km\,h^{-1}}$ relative to the ground (G).

 a The wind is blowing from east to west at $200\,\mathrm{km\,h^{-1}}$ relative to the ground. Calculate the plane's velocity relative to the wind (W).

 _____

 b The wind is blowing from west to east at $200\,\mathrm{km\,h^{-1}}$ relative to the ground. Deduce the aeroplane's velocity relative to the wind.

 _____

 _____

 c If the wind is blowing from south to north at $200\,\mathrm{km\,h^{-1}}$ relative to the ground, but the aeroplane still wants to go directly west, does the aeroplane's motion *relative to the wind* need a component to the north or south? If so, how large must that component be? And in what direction?

 _____

 _____

 d **i** The wind is now blowing from south to north at $200\,\mathrm{km\,h^{-1}}$ relative to the ground, but the aeroplane is still going directly west at $700\,\mathrm{km\,h^{-1}}$ relative to the *ground.* Calculate the magnitude of the aeroplane's velocity relative to the *wind.*

 _____

 _____

 _____

 ii We now know what *speed* the aeroplane must have relative to the wind. Determine the *direction* the pilot must point the plane in if it is to go west at $700\,\mathrm{km\,h^{-1}}$. Express this as an angle relative to west.

 _____

 _____

 _____

2 Diana is trying to row across a river. As seen from the bank, the river is flowing from left to right at $0.80\,\mathrm{m\,s^{-1}}$. If Diana rows perpendicular to the current, the combination of her rowing and the current will cause her to reach the other side some distance downstream of where she starts.

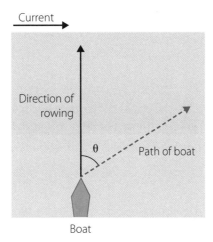

a Diana points the boat perpendicular to the bank and rows forward at $0.50\,\mathrm{m\,s^{-1}}$, but gets carried sideways by the current. Calculate her total speed relative to the *bank*.

_____

_____

b Determine the value of θ.

_____

c If the river is 50.0 m wide, how far downstream will Diana go before reaching the other bank?

_____

_____

d If the boat is to reach the other side opposite to where it started, Diana must point the boat somewhat upstream. Deduce the magnitude of the upstream component of its velocity.

_____

_____

e Sketch the situation.

f The river is 50.0 m wide. Diana wants to get to the other side in 50.0 s. Calculate the magnitude of the component of her velocity perpendicular to the bank.

_____

_____

g Using the information and your answers to parts d and e, what is the velocity of the boat *relative to the water*?

_____

_____

_____

_____

3 Four cars are driving across the desert. The black car (B) is going north at $100\,km\,h^{-1}$. The green car (G) goes south at $80\,km\,h^{-1}$. The orange car (O) goes south-east at $80\,km\,h^{-1}$ and the red car (R) goes east at $80\,km\,h^{-1}$.

a Sketch a velocity vector diagram showing the speeds and directions of the four cars.

b What is the velocity of the green car relative to the black car?

_____

_____

_____

c What is the velocity of the red car relative to the black car?

_____

_____

_____

_____

d What is the velocity of the orange car relative to the black car?

_____

_____

_____

_____

_____

Extension

e If we take θ as the angle away from north in the clockwise direction, find a general expression for the *speed* relative to the black car of a car travelling in direction θ at $80\,\text{km h}^{-1}$. Call it the yellow car (Y).

_____

_____

_____

_____

_____

4 Two football players are running. Yolanda, in the yellow jersey (Y), is running north at $5.0\,\text{m s}^{-1}$. Brendan, in the blue (B), is running S20°E at $6.0\,\text{m s}^{-1}$.

a Sketch a vector diagram of the speeds and directions of the players.

b What is Yolanda's velocity relative to Brendan?

_____

_____

_____

_____

c The ball is rolling west at $4.0\,\text{m s}^{-1}$. Which player is moving faster relative to the ball?

_____

_____

HINT

Because the requested answer here is *qualitative*, do you need to do all the maths? Would a scale diagram work?

Module one: Checking understanding

Circle the correct answer for questions 1–6.

1 Which of these quantities is a vector?

   A Time

   B Displacement

   C Distance

   D Speed

2 Using a graph of velocity plotted against time (an example is shown below), how can we get the total displacement?

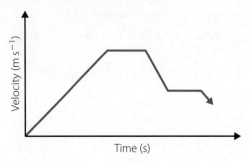

   A We can't, there is not enough information.

   B By finding the average slope.

   C By multiplying the final velocity by the final time.

   D By working out the area under the graph.

3 The acceleration due to gravity near the surface of Mars is $3.71\,\mathrm{m\,s^{-2}}$. If a ball is dropped from a height of $2.00\,\mathrm{m}$, how long will it take to fall to the ground?

   A $1.04\,\mathrm{s}$

   B $0.64\,\mathrm{s}$

   C $1.69\,\mathrm{s}$

   D $2.64\,\mathrm{s}$

4 I started at my front gate and ran $800\,\mathrm{m}$ north and then turned and ran $600\,\mathrm{m}$ west. How far am I from my front gate?

   A $1400\,\mathrm{m}$

   B $200\,\mathrm{m}$

   C $1000\,\mathrm{m}$

   D $980\,\mathrm{m}$

5 Here are two vectors, $\vec{A}$ and $\vec{B}$, represented graphically.

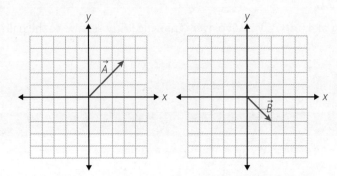

Which diagram correctly represents the vector subtraction $\vec{R} = \vec{A} - \vec{B}$?

A

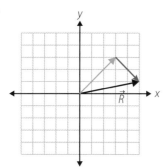

B

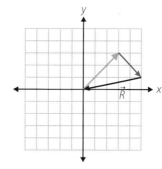

C

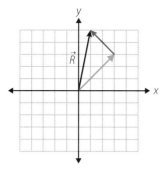

D

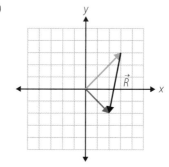

6 Two children are in the back seat of a car. The car is driving east at $60.0 \, \text{km h}^{-1}$. The child seated on the left throws a ball across the car at the child seated on the right. The ball travels $1.50 \, \text{m}$ in $0.300 \, \text{s}$. What is the magnitude of the velocity of the ball relative to the ground?

A $^14.8 \, \text{m s}^{-1}$

B $18.9 \, \text{m s}^{-1}$

C $60 \, \text{km h}^{-1}$

D $65 \, \text{km h}^{-1}$

7 To reach escape velocity, a rocket must be travelling at $11.0 \, \text{km s}^{-1}$ when it is $200 \, \text{km}$ above Earth's surface. If it is moving at $4.00 \, \text{km s}^{-1}$ when it is $100 \, \text{km}$ up, what must its average acceleration be over the next $100 \, \text{km}$ if it is to reach escape velocity?

8 Mahmud is riding a flying fox across a river. The cable of the flying fox is perpendicular to the river bank. A duck on the water is floating with the current and is moving from Mahmud's left to right at $2.0 \, \text{m s}^{-1}$. The flying fox is moving at $5.0 \, \text{m s}^{-1}$.

a Sketch the situation, including vectors for Mahmud's velocity and that of the duck.

b What is Mahmud's velocity relative to the duck?

_____

_____

_____

_____

_____

_____

_____

_____

_____

MODULE TWO »
DYNAMICS

Reviewing prior knowledge

1 In each of the following situations, determine whether the object is acted upon by a balanced force or by an unbalanced force.

   a An elevator that is stationary on level 7 of a building   _____

   b A bicycle that gets faster as it goes downhill   _____

   c A bus being driven along a straight highway at a constant $50\ km\,h^{-1}$ _____

   d A ball that slows down when rolling across flat ground   _____

   e A motorcycle that goes around a corner at a constant speed   _____

2 In each of the following situations, describe the source of the frictional force that is acting and one way it could be reduced.

   a A boy struggles to pull a heavy wooden box across a bumpy dirt road. (Force on the box)

   _____

   _____

   _____

   b A skydiver slows down after opening her parachute.

   _____

   _____

   _____

   c A coach finds it hard to squeeze the end of the pump into the narrow rubber valve on a soccer ball.

   _____

   _____

   _____

3 A cyclist notices that after a long difficult downhill section the disc brakes of her mountain bike are extremely hot. What is the source of this heat?

_____

_____

_____

_____

_____

4   Describe three situations in which a student might show her grandfather three different types of non-contact forces in action in a normal family home.

_____

_____

_____

_____

_____

5   How does the mass of an object differ from the weight of an object?

_____

_____

_____

_____

6   A modern car features several energy conversion devices. Complete the table below to indicate the energy conversion that the device enables. The first one has been done for you.

Conversion event	Converts energy from	Converts energy to
Headlights are turned on	Electrical energy	Light energy
Combustion engine is operating to drive car		
Car radio is turned on		
Boot opens automatically		
Car heater is turned on		

7   Choose one of the energy conversions from question 6 to describe how some of the 'energy converted from' is wasted by being turned into an unwanted form in the conversion process.

_____

_____

_____

_____

Forces

 WS 4.1 Field forces, contact forces and Newton's laws of motion

STUDENT BOOK
Pages 89–94

LEARNING GOALS

Apply Newton's laws of motion, in static and dynamic situations, for contact forces and field forces

We have studied kinematics, which allows us to describe *how* objects move. But to understand *why* objects move as they do, we need to understand forces.

When objects interact, they do so via forces. Newton's laws of motion tell us how forces affect the motion of objects.

1 You hold an apple in your hand. Is the force you exert on the apple a field force or a contact force? Why?

_____

2 The contact force has two components. What are these components called, and what are their directions?

_____

_____

3 Name three field forces and give examples of when each is acting.

_____

_____

_____

4 State Newton's first law and explain what it means.

_____

_____

_____

_____

_____

5 A ball rolling along the floor gradually slows down, until it comes to a stop. How would Newton explain this behaviour?

_____

_____

_____

_____

6 Write Newton's second law as an equation and in words. Define all the terms in the equation.

_____

_____

_____

7 Using Newton's second law, define what is meant by a *static* situation and what is meant by a *dynamic* situation.

_____

_____

8 Write Newton's third law as an equation and in words.

_____

_____

9 Narelle asks Emma to push her while she sits in her billy cart. But Emma says that she cannot, because by Newton's third law, whatever force she applies to Narelle, Narelle will apply an equal and opposite force to her and hence she can never make Narelle move. What is wrong with Emma's argument?

_____

_____

_____

10 Narelle is pushing Emma in the billy cart to demonstrate that it can be done. When Narelle first starts pushing and Emma is accelerating, is the force by Narelle on Emma greater than, equal to, or smaller than the force that Emma exerts on Narelle? Justify your answer.

_____

_____

11 Emma tells Narelle to stop pushing her so fast, so Narelle slows down and pushes so that Emma moves at a constant speed. Is the force by Narelle on Emma greater than, equal to, or smaller than the force Emma exerts on Narelle now? Justify your answer.

_____

_____

12 For each of the following forces, give the Newton's third law force pair.

a The gravitational force of Earth on the Moon

_____

b The gravitational force of Earth on this book

_____

c The force of a magnet on a fridge door

_____

d The friction force of a car's tyre on the road

_____

e The normal force of the floor on you when you stand up

_____

Use algebraic and vectorial methods to analyse concepts of net force and equilibrium in one-dimensional and two-dimensional situations

When an object is in equilibrium it means that the total, or net, force acting on it is equal to zero. But to find the total force we must remember that forces are vectors. To find the net force acting on an object we must add the forces in each direction. For an object to be in equilibrium, the total force in *all* directions must be zero.

1 Give an example of when an object is in equilibrium but not at rest.

_____

_____

_____

2 Nick and Ed are having a tug of war with a toy unicorn. Nick pulls it to the left with a force of 100 N and Ed pulls to the right with a force of 120 N.

a What is the net force on the unicorn? Give magnitude and direction.

_____

b Which way does the unicorn accelerate?

_____

c Nick now increases the force he applies to 120 N left. What is the acceleration of the unicorn now?

_____

d Is the unicorn now in equilibrium? Explain your answer.

_____

e Is the unicorn now stationary? Explain your answer.

_____

3 Mai is taking her father breakfast in bed. She puts a cup of coffee on a tray and lifts the tray up. For each of the following situations, is the normal force of the tray on the cup greater than, equal to, or smaller than the gravitational force acting on the cup (the weight of the cup)? Justify your answers.

a Mai lifts the tray, accelerating it upwards.

_____

_____

_____

b Mai carries the tray at constant height and constant velocity.

_____

_____

_____

c Mai lowers the tray, allowing it to accelerating it gently downwards.

_____

_____

_____

4 Draw a force diagram for each of the following situations, and state the direction of the net force, or if the net force is zero. Show all forces acting on James, both horizontal and vertical, on your diagram.

a James sits cross-legged on the floor.

b James is sliding to the left along the floor in his socks, slowing down.

c James is running to the right, speeding up as he goes.

d James has just jumped off a diving board and is falling towards the water below.

5 Two children are pushing on a trolley. Manus pushes it north with a force of 150 N and Liam pushes it west with a force of 180 N.

a Draw a scale diagram to show the forces and measure the magnitude of the net force and its angle (from west to north) from your diagram.

> **HINT**
>
> Choose a sensible scale given the space available, for example 1 cm = 20 N. You will need a protractor to measure the angle.

b Use trigonometry to calculate the net force on the trolley and give its direction.

_____

_____

c Do your two answers agree? Which method is more precise, and why?

_____

_____

6 Anika is fishing for yabbies in a dam using a small piece of meat tied to a string as bait. A yabby has grabbed hold of the bait and is pulling horizontally with a force of 16 N. Anika pulls directly upwards with a force of 20 N.

   a What is the net force on the bait? Give magnitude and direction, and draw a diagram showing the forces and the net force.

---

   b A second yabby grabs hold of the bait. What force must it exert on the bait if the bait is to be in equilibrium? Give magnitude and direction.

---

   c If the second yabby exerts this force on the bait, what force does the bait exert on this second yabby? Give magnitude and direction.

---

7 Two forces act on an object, a vertical force of 450 N and a horizontal force of 270 N. Calculate the net force acting on the object and give its direction to the horizontal. A diagram may be useful.

---

8 Two perpendicular forces, $\vec{F}_A$ and $\vec{F}_B$, act on an object, creating a net force of $\vec{F}_{net} = 250$ N.

   a If $\vec{F}_A = 200$ N, what is the magnitude of $\vec{F}_B$?

---

   b Draw a diagram showing $\vec{F}_{net}$, $\vec{F}_A$ and $\vec{F}_B$.

   c Calculate the angles between $\vec{F}_{net}$ and $\vec{F}_A$ and between $\vec{F}_{net}$ and $\vec{F}_B$, and mark these on your diagram.

Finding resultant forces using force components

LEARNING GOALS

Solve force problems using vector resolution techniques

Forces are vectors. When adding forces in two dimensions, we need to resolve the forces into perpendicular components and then add the components.

1 A force vector can be written $\vec{F} = \vec{F}_x + \vec{F}_y$. If the angle between $\vec{F}$ and $\vec{F}_x$ is θ, write expressions for $\vec{F}_x$ and $\vec{F}_y$ in terms of $\vec{F}$ and θ.

_____

2 A force has a magnitude of 300 N. What is the maximum possible value of its y component? Under what condition does the y component take this maximum value?

_____

_____

_____

_____

3 A force is at 35° to the horizontal, and has magnitude 550 N. Calculate the x and y components of the force.

_____

_____

_____

4 A triangular shade sail is suspended by three ropes so it hangs horizontally, in equilibrium. Considering only the forces in the horizontal plane, the forces exerted by ropes one and two are:

$\vec{F}_1 = \vec{F}_{1,x} + \vec{F}_{1,y} = 100\,\text{N east} + 100\,\text{N north}$

$\vec{F}_2 = \vec{F}_{2,x} + \vec{F}_{2,y} = 50\,\text{N west} + 100\,\text{N south}$

a Calculate the force exerted by rope 3, $\vec{F}_3$. Consider only the horizontal forces.

_____

_____

_____

_____

_____

b Draw a diagram showing the forces acting on the shade sail.

5 For each of the following situations, calculate the net force acting on the block. Give the *x* and *y* components of the force, its magnitude and its angle to the positive *x*-axis. Draw the net force on the diagram.

a

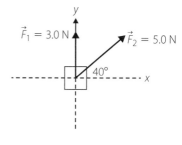

b

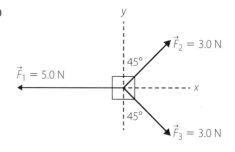

c

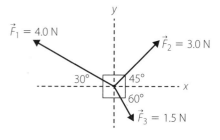

6 A force is applied by a rope to a fallen tree.

a If the magnitude of the force is 3500N and the vertical component for the force is 1650N, what is the angle the rope makes with the horizontal?

_____

_____

b Calculate the horizontal component of the force.

_____

_____

c The gravitational force acting on the fallen tree is given by $\vec{F}_g = m\vec{g}$, where m is the mass of the tree, $m = 450\,\text{kg}$, and $\vec{g}$ is the acceleration due to gravity, $\vec{g} = 9.8\,\text{m s}^{-2}$ down. Using this information, calculate the gravitational force on the tree, and give its magnitude and direction.

_____

_____

d Draw a force diagram for the tree. Do not forget the contact force due to the ground.

e Assuming the force due to the rope is not enough to start the tree moving, calculate the normal and friction forces exerted on the tree by the ground.

_____

_____

_____

f Calculate the total contact force by the ground on the tree, and give its magnitude and direction.

_____

_____

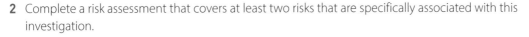
LEARNING GOALS

Investigate the motion of objects on an inclined plane

INVESTIGATION

Measuring the frictional force using an inclined plane

AIM

To measure the maximum static friction force between a board and several different objects

RESEARCH QUESTION

1 Write a research question for this experiment.

RISK ASSESSMENT

2 Complete a risk assessment that covers at least two risks that are specifically associated with this investigation.

RISK
ASSESSMENT

What are the risks in doing this investigation?	How can you manage these risks to stay safe?

MATERIALS

- Short length of timber, such as a cutting board   • Protractor   • Kitchen scales
- Small objects, such as erasers, pencil cases, notebooks. Choose items with different textures

METHOD

1 With the board horizontal, place a small object on the board.
2 Slowly tilt the board.
3 When the object on the board just starts to slide, record the angle of the board to the horizontal.
4 Repeat steps 1 to 3 twice, so you have three measurements of the angle for each object.
5 Replace the object with a different object and repeat steps 1 to 4.
6 Measure the mass of each item using the kitchen scales.
7 Record your results in the first five columns of the table below.

Item	Angle 1 (°)	Angle 2 (°)	Angle 3 (°)	Average angle (°)	Mass (kg)	Maximum friction force (N)

3 Draw a force diagram showing the forces acting on the object on the inclined board. Mark the angle between the board and the horizontal as **θ**.

4 Now redraw your force diagram, with the forces broken into components parallel to the board and perpendicular to the board.

5 Using your force diagram, derive an equation for the friction force on the object just before it starts to slide.

_____

_____

_____

_____

_____

_____

6 Using your measurements in the table above, calculate the maximum frictional force acting on each object due to the board.

_____

_____

_____

_____

_____

_____

7 Write these values into the table.

8 Consider the following questions and discuss them with your group. Use the space below to make notes.

 a How does the maximum frictional force depend on the texture of an object?

 b For objects with similar textures, did the frictional force depend on the mass of an object?

 c If you were going to carry out this investigation again, what changes would you make to ensure a more valid/reliable investigation?

CONCLUSION

9 Give the answer to your research question.

INQUIRY QUESTION: HOW CAN THE MOTION OF OBJECTS BE EXPLAINED AND ANALYSED?

WS **5.1** Applying Newton's laws of motion

STUDENT BOOK
Pages 117–26

LEARNING GOALS

Analyse forces, in both static and dynamic everyday situations, using Newton's Laws

Newton's first two laws of motion tell us how forces affect the motion of objects. The first law says that if there is no net force, then there is no change in the state of motion. The second law tells us that the acceleration of an object is determined by the net force acting on the object. When we calculate the net force acting on an object, we need to consider all the forces that are acting – both field forces and contact forces. The contact force has two components – the normal force and the friction force. The friction force is important because nearly all our motion in everyday situations relies on friction.

1 An object is at rest. Does this mean that there are no forces acting on it? Explain your answer.

_____

_____

_____

_____

2 An object is moving at constant velocity. Does this mean there must be a force in the direction of motion? Explain your answer.

_____

_____

_____

_____

3 An object is accelerating. Does this mean there must be a force in the direction of motion? Explain your answer.

_____

_____

_____

_____

4 Why is the static friction force necessary for a car to move? Use a diagram to help explain your answer, and show where the static friction force acts.

_____

_____

_____

_____

5 A car has a mass of 1500 kg, and the coefficient of static friction between the car's tyres and the road is $\mu_s = 0.75$. What are the minimum and maximum values of the static frictional force on the car when it is on a flat road? Under what conditions do these forces occur?

_____

_____

_____

_____

6 For the car in question **5**, what is the maximum acceleration of the car on a flat road? (Assume the power of the engine is not the limiting factor.)

_____

_____

7 What is the minimum stopping distance for this car when it is travelling at a speed of $100 \, \text{km h}^{-1}$?

_____

_____

_____

8 Explain why it is important to slow down on a wet or slippery road. What happens to the minimum stopping distance?

_____

_____

_____

9 The coefficient of kinetic friction is generally lower than the coefficient of static friction. Based on this, explain why the minimum stopping distance is greater when a car is skidding.

_____

_____

_____

10 Draw a force diagram for a car that is travelling at a constant high speed. Identify each force acting on the car by stating what is exerting the force and where on the car it acts.

_____

_____

_____

_____

11 Referring to your diagram from question **10**, explain why the engine needs to continue running for the car to maintain a constant velocity on a flat road.

_____

_____

_____

_____

12 Neil is sliding across his kitchen floor in his socks. The coefficient of kinetic friction between Neil's socks and the floor is 0.20. If Neil is sliding at a speed of $3.5\,\mathrm{m\,s^{-1}}$, at what speed will he collide with the wall 3.0 m away?

_____

_____

_____

_____

13 If Neil was sliding across the kitchen in his socks just like this on a Moon base, where the acceleration due to gravity is only $1.6\,\mathrm{m\,s^{-2}}$, at what speed would he hit the wall?

_____

_____

Analysing the acceleration of objects subject to a constant net force

STUDENT BOOK
Pages 126–32

Analyse graphically, vectorially, mathematically and qualitatively the acceleration of an object subject to a constant net force using newton's second law

When an object is subjected to a net force, it has an acceleration in the direction of the net force. If the force is constant, then the acceleration is also constant. The magnitude of the acceleration can be calculated from Newton's second law: $\vec{F} = m\vec{a}$.

1 Describe how an object's velocity varies with time when it is subject to a constant net force.

_____

_____

2 In which of the following situations is the car subject to a constant net force? Explain why or why not.

a A car is travelling at constant speed on a flat, straight road.

_____

_____

b A car is travelling at constant speed on a flat road, going around a corner.

_____

_____

c A car is travelling with constant acceleration in a straight line up a hill.

_____

_____

d A car is accelerating from rest, with acceleration decreasing steadily as it approaches the speed limit.

_____

_____

3 Kristy drops an economics book out a window. Sketch graphs of the following as a function of time while the book is falling. Ignore air resistance.

a the acceleration of the book

b the velocity of the book

c the position of the book

4 A car is travelling at $10\,\text{m}\,\text{s}^{-1}$ on a straight, flat road when the driver applies the brakes. Take the direction of the car's motion to be the positive x direction, and its position when the brakes are first applied to be $x = 0$. The car has a mass of $1200\,\text{kg}$, and the coefficient of friction between the road and the tyres is 0.80.

   a Calculate the maximum friction force on the car by the road.

_____

_____

   b Assuming the driver brakes as hard as possible without skidding, calculate the acceleration of the car.

_____

_____

   c Calculate the time taken for the car to come to a stop.

_____

_____

   d Calculate the distance taken for the car to come to a stop.

_____

_____

_____

   e Calculate the distance travelled by the car in the first half of the time it takes to stop.

_____

_____

_____

   f On the axes below, draw *to scale* graphs of the acceleration, velocity and position of the car from when it starts braking to when it comes to a stop.

     i acceleration

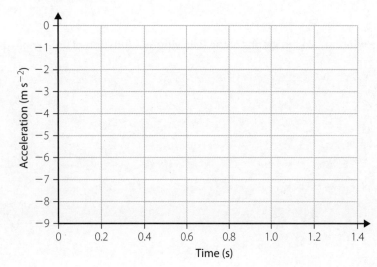

ii velocity

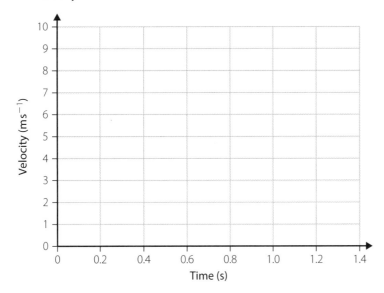

iii position

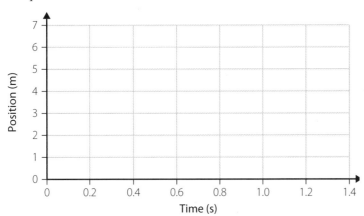

5 Air resistance varies with an object's speed, getting larger as the object goes faster. Explain why we cannot use the kinematics equations $v = u + at$ and $s = ut + \dfrac{1}{2}at^2$ to model the motion of a falling object when air resistance is not negligible.

_____

_____

_____

6 A net force of $\vec{F}_{net} = 250\,N$ pointing at an angle of $45°$ above the horizontal acts on an object. The object has a mass of $20\,kg$.

  a What is the direction and magnitude of the object's acceleration?

_____

_____

b Draw vector diagrams showing:

    i the net force on the object

    ii the acceleration of the object.

7 The box shown in each of the diagrams below has a mass of 1.5 kg. Calculate the acceleration of the box in each case and draw a vector to scale of the acceleration on a separate diagram. Use the scale $1 \, \text{m s}^{-2} = 1 \, \text{cm}$.

a

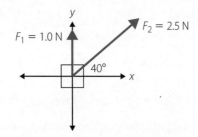

_____

_____

_____

_____

_____

b

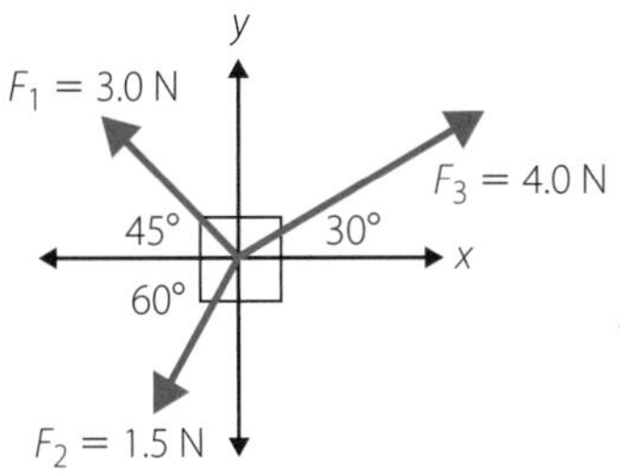

$F_1 = 3.0\,\text{N}$
$F_3 = 4.0\,\text{N}$
$45°$ $30°$
$60°$
$F_2 = 1.5\,\text{N}$

----

8 Marcus is sitting on a blanket on the floor and Laurence is dragging him along by pulling on the blanket. Marcus has a mass of 35 kg, and the coefficient of kinetic friction between the blanket and the floor is 0.15. The mass of the blanket is negligible compared to the mass of Marcus. If Laurence pulls with a force of 120 N at an angle of 27° to the horizontal, what is Marcus's acceleration?

Apply principles of conservation of mechanical energy

Energy is one of the central concepts in physics. There are two types of energy: kinetic and potential. Energies and forces are related: when a force acts on an object, it does work and energy is transferred.

1 Define 'kinetic energy'.

_____

_____

2 Write the equation for kinetic energy and define all the symbols used.

_____

_____

3 Is it possible for kinetic energy to be negative? Why, or why not? If it can be negative, give an example.

_____

_____

_____

4 A 1500 kg car is travelling at 100 km h^{-1}.

  a Calculate the kinetic energy of the car.

_____

_____

_____

  b If the car has a kinetic energy of 210 kJ, what speed is it travelling at? Give your answer in km h^{-1}.

_____

_____

_____

5 Define 'potential energy'.

_____

_____

6 Write the equation for the potential energy of an object in Earth's gravitational field, close to Earth's surface. Define all the symbols used.

_____

_____

_____

7 Is it possible for potential energy to be negative? Why, or why not? If it can be negative, give an example.

_____

_____

_____

8 Kristy holds a 1.5 kg economics book out of a window, 3.5 m above ground.

    a Relative to the ground, what is the gravitational potential energy of the book?

    _____

    b Relative to Ann looking out a window on the floor above, 2.4 m higher than the book, what is the gravitational potential energy of the book?

    _____

    _____

9 Define 'mechanical energy'. What types of energy are _not_ included in mechanical energy?

_____

_____

_____

10 Under what conditions is the mechanical energy of a system conserved?

_____

_____

11 Define 'work' and give its units and the equation used to calculate it.

_____

_____

_____

12 Draw diagrams below showing force and displacement vectors for when the work done by a force is:

    a maximum magnitude and positive

    b maximum magnitude and negative

    c zero.

13 The graph below shows the net force acting on an object, in the direction of its displacement, as a function of its displacement. Calculate the total work done on the object by the net force.

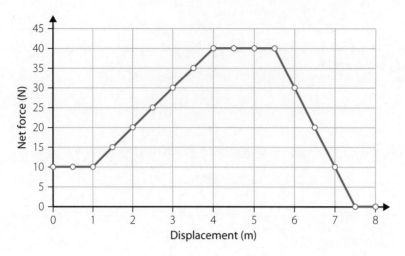

_____

_____

_____

_____

_____

14 A 1800 kg car is being towed using a tow rope tied to the car. The car is on a flat horizontal road and the tow rope makes an angle of 25° to the horizontal.

a If the tow rope applies a net force of 500 N to the car, how much work does the force applied by the tow rope do on the car for every kilometre the car is towed?

_____

_____

_____

_____

b If the car started at rest, at what speed is it travelling after 1 km, if no other forces are acting in the horizontal direction?

_____

_____

_____

15 A book slides across a table, slowing down due to friction as it goes. Is friction doing positive or negative work in this case? Explain your answer.

_____

_____

16 Is it possible for friction to do positive work? Explain your answer.

_____

_____

_____

17 Kristy drops a 1.5 kg economics book out of a window 3.5 m above the ground. What is the speed of the book when it hits the ground? Use conservation of mechanical energy to calculate the answer. Ignore air resistance.

_____

_____

_____

_____

18 Alex slides his 1.5 kg mathematics book down a slide. The slide is straight, 12 m long and makes an angle of 35° with the horizontal.

    a Ignoring friction, with what speed does the book hit the ground?

_____

_____

_____

    b If the slide started and ended at the same points as in part **a** but was 30 m long and went around in a spiral, with what speed would the book hit the ground? Ignore friction.

_____

_____

    c In reality, friction acts between the book and the slide. In which case, **a** or **b**, would the book really hit the ground faster? Explain your answer in terms of work done on the book.

_____

_____

_____

Understand the relationship between power, time and energy transformed or transferred

Apply this relationship to solving mechanical problems

Power is the rate at which energy is transformed from one form to another, or transferred from one object or system to another. Energy is transferred when work is done by a force, so in dynamics power is the rate at which work is done by a force.

1 Define 'power' and give its units.

_____

_____

2 A kettle is rated at 1.5 kW. How much energy can it convert from electrical potential energy to thermal energy per minute?

_____

_____

3 A car can accelerate from 0 to 100 km h^{-1} in 11 s. If the car has a mass of 1200 kg, what power is required to achieve this acceleration?

_____

_____

_____

4 Neil, mass 48 kg, is sliding across the kitchen floor in his socks. His initial speed is 2.0 m s^{-1}. The kinetic friction force does work on him at a rate of 38 W. How long does it take for Neil to slide to a halt?

_____

_____

_____

5 Kristy drops an economics book out a second-floor window. Sketch a graph of the power transferred by the gravitational field to the book as a function of time. Explain the shape of your graph.

_____

_____

_____

6 A lift is being raised by a motor. The combined mass of the lift and the people in it is 750 kg and it moves upwards at constant speed a distance of 10 m.

a What force must the motor provide to move the lift upwards at constant speed?

b What is the change in gravitational potential energy of the lift (including the people inside) as it rises through 10 m?

c How much work has the motor done on the lift?

d If the lift takes 12 s to travel the 10 m distance, what is the power output of the motor?

e If the power output of the motor was 10 kW, at what speed could the lift travel upwards?

7 A ball of mass 250 g is rolled across the floor. The ball is initially moving at $1.5 \, \text{m s}^{-1}$. It takes 5.0 m for the ball to come to a rest.

a Calculate the average force of rolling friction on the ball, using an energy approach.

b Calculate the average power of the rolling friction force on the ball.

8 A 1200 kg car is travelling at a constant $80 \, \text{km h}^{-1}$ on a straight, flat road. The force of air resistance on the car is 350 N.

a Ignoring rolling resistance, what is the frictional force of the road on the car's tyres?

b Calculate the rate at which the static friction force does work (the power) on the car.

c The car slows down such that the air resistance drops to 150 N and only 2.1 kW power is required to keep the car at steady speed. What speed is the car doing now? Give your answer in km h^{-1}.

_____

_____

_____

9 Marcus is sitting on a blanket on the floor and Laurence is dragging him along by pulling on the blanket. Marcus has a mass of 35 kg and the mass of the blanket is negligible compared to his mass. Laurence pulls with a force of 150 N at an angle of 27° to the horizontal.

a Find the force applied by Laurence in the direction of Marcus's motion.

_____

_____

_____

b What is Marcus's kinetic energy after 1 s? Assume that he starts at rest, and ignore friction.

_____

_____

_____

_____

c At what rate (power) is Laurence doing work on Marcus?

_____

_____

6 Momentum, energy and simple systems

 Investigating one-dimensional interactions in a closed system

STUDENT BOOK
Pages 153–60

LEARNING GOALS

Analyse a one-dimensional collision through a first-hand investigation

Momentum is an important quantity in physics and is useful for analysing motion. The momentum, $\vec{p}$, of an object is a vector quantity given by $\vec{p} = m\vec{v}$ where m is the object's mass and $\vec{v}$ is its velocity. The momentum has the same direction as the velocity. The total momentum of an isolated system is constant: momentum is a conserved quantity.

INVESTIGATION

One-dimensional collisions

AIM

To investigate collisions between similar and differently sized objects in a simple one-dimensional system

RESEARCH QUESTION

1   Write a research question for this investigation.

_____

_____

_____

_____

RISK ASSESSMENT

2   Complete a risk assessment that covers at least two risks that are specifically associated
with this investigation.

RISK ASSESSMENT

What are the risks in doing this experiment?	How can you manage these risks to stay safe?

- 2 rulers   • 3 or more marbles, in 2 different sizes   • Stopwatch and video camera *or* 2 smart phones
- Weighing scale (electronic balance), such as good quality kitchen scale

METHOD

1 Place the two rulers parallel, about 1 cm apart, on a flat surface to make a track for the marbles.
2 Place the stopwatch (or smart phone displaying stopwatch) next to the track.
3 Weigh and record the mass of each marble.
4 Place one marble about halfway along the track.
5 Place a second marble of the same size at the start of the track.
6 Position the video camera (or smart phone in video mode) so that it has a clear view of the entire track and the stopwatch. You need to be able to read the stopwatch and the position of the object in the picture.
7 Start the stopwatch and video camera.
8 Flick the marble at the start of the track along the track towards the first marble.
9 View the recording and note the times and positions of each marble at two times shortly before the collision (times and positions 1 and 2 in the results table) and at two times shortly after the collision (times and positions 3 and 4 in the results table).
10 Repeat steps 3 to 9 with different sized marbles. Try flicking a large marble at a small marble and vice versa.

RESULTS AND ANALYSIS

Experiment one: same sized marbles

3 Write in words what you observed.

_____

_____

_____

_____

4 Complete the table of results using your video recording.
   Use the times and positions recorded to calculate the speed of each marble just before and just after the collision. Use the masses of the marbles and the calculated velocities to calculate the momentum and kinetic energy of each marble and of the whole system before and after the collision. Note that you need to decide which direction is positive and which is negative.

5 Write down the equations that you will use to calculate velocity, momentum and kinetic energy.

_____

_____

_____

_____

_____

_____

_____

	Marble 1	Marble 2	System (Marble 1 + Marble 2)
Mass (kg)			
Time 1 (s)			
Position 1 (m) (position at time 1)			
Time 2 (s)			
Position 2 (m) (position at time 2)			
Velocity before collision (m s⁻¹)			
Momentum before collision (kg m s⁻¹)			
Kinetic energy before the collision (J)			
Time 3 (s)			
Position 3 (m)			
Time 4 (s)			
Position 4 (m)			
Velocity after collision (m s⁻¹)			
Momentum after collision (kg m s⁻¹)			
Kinetic energy after the collision (J)			

Experiment two: different sized marbles, small marble initially stationary

6   Write in words what you observed.

_____

_____

_____

_____

7   Complete the table of results using your video recording.

	Marble 1	Marble 2	System (Marble 1 + Marble 2)
Mass (kg)			
Time 1 (s)			
Position 1 (m)			
Time 2 (s)			
Position 2 (m)			
Velocity before collision (m s⁻¹)			
Momentum before collision (kg m s⁻¹)			
Kinetic energy before the collision (J)			
Time 3 (s)			
Position 3 (m)			
Time 4 (s)			
Position 4 (m)			
Velocity after collision (m s⁻¹)			
Momentum after collision (kg m s⁻¹)			
Kinetic energy after the collision (J)			

Experiment three: different sized marbles, large marble initially stationary

8 Write in words what you observed.

_____

_____

_____

_____

9 Complete the table of results using your video recording.

	Marble 1	Marble 2	System (Marble 1 + Marble 2)
Mass (kg)			
Time 1 (s)			
Position 1 (m)			
Time 2 (s)			
Position 2 (m)			
Velocity before collision (m s^{-1})			
Momentum before collision (kg m s^{-1})			
Kinetic energy before the collision (J)			
Time 3 (s)			
Position 3 (m)			
Time 4 (s)			
Position 4 (m)			
Velocity after collision (m s^{-1})			
Momentum after collision (kg m s^{-1})			
Kinetic energy after the collision (J)			

DISCUSSION

10 Consider the following questions and discuss them with your group. Use the space below to make notes.

   a How did the momentum of each marble, and of the system of two marbles, change as a result of their interaction?

   b How did the kinetic energy of each marble, and of the system of two marbles, change as a result of their interaction?

   c Why is it important to use a short time period close to the collision both before and after to calculate the velocity and hence momentum of each marble?

   d How well did the marbles in this experiment approximate an isolated system?

   e If you were going to carry out this investigation again, what changes would you make to ensure a more reliable investigation? How can you ensure the investigation is valid?

   f How could you extend this investigation?

_____

_____

_____

_____

_____

_____

_____

_____

_____

_____

_____

_____

_____

_____

_____

_____

_____

CONCLUSION

11 Have you answered your research question? Write a suitable conclusion to your investigation, making sure that you are relating it back to your research question.

_____

_____

_____

_____

_____

_____

_____

_____

_____

LEARNING GOALS

Analyse elastic collisions through the application of conservation laws for momentum and energy

In any interaction, momentum is conserved. This is a direct result of Newton's third law, which says that, in any interaction, the interacting objects exert equal and opposite forces on each other: $\vec{F}_{A \text{ on } B} = -\vec{F}_{B \text{ on } A}$.

Over the time taken for the interaction to occur, Δt, each object's momentum changes by $\Delta \vec{p} = \vec{F} \Delta t$
so $\Delta \vec{p}_B = \vec{F}_{A \text{ on } B} \Delta t = -\vec{F}_{B \text{ on } A} \Delta t = -\Delta \vec{p}_A$. This is true for any interaction, of any force type, between two objects.
In an elastic collision, kinetic energy is also conserved; no energy is converted to any other type.

1 A 10 g marble is moving at $20 \, \text{cm s}^{-1}$. Calculate the momentum of the marble.

2 A 1500 kg car has a momentum of $23\,000 \, \text{kg m s}^{-1}$. Calculate its speed. Give your answer in km h^{-1}.

3 Show that the kinetic energy of an object can be written in terms of its momentum as

$$E_k = \frac{p^2}{2m}$$

4 A tennis ball has a mass of 58 g and a momentum of $0.46 \, \text{kg m s}^{-1}$. Calculate the kinetic energy of the tennis ball.

5 A tennis ball has a mass of 58 g and a kinetic energy of 0.75 J. Calculate the momentum of the tennis ball.

6 A cat and a dog have the same kinetic energy. The dog has twice the mass of the cat.

   a What is the ratio of the cat's velocity to the dog's velocity?

   b Which has the larger momentum? Explain your answer.

7 Sally drops a 0.1 kg super-ball from a height of 1.3 m on to the floor, where it makes an elastic collision.

   a To what height will the ball rebound if the collision is perfectly elastic (ignoring air resistance)? Explain your answer and state any assumptions or approximations you make.

   b Ignoring air resistance, calculate the speed with which the ball hits the floor.

   c What is the momentum change of the ball as a result of the collision?

   d What is the kinetic energy of the ball just before it hits the floor?

   e What is the change in kinetic energy of the ball as a result of the collision?

   f What is the momentum change of Earth as a result of the collision?

   g If we take Earth as stationary before the collision, what is its speed after the collision? The mass of Earth is 6.0×10^{24} kg.

h What is the kinetic energy of Earth after the collision?

_____

i Using the calculations above, justify the approximation that Earth's speed, momentum and kinetic energy do not change in these types of collisions.

_____

_____

_____

8 Mika flicks a marble at a wall. It hits the wall and bounces directly backwards, with the same speed but in the opposite direction.

a Is kinetic energy conserved in this collision? Explain your answer.

_____

_____

b Is momentum conserved in this collision? Explain your answer.

_____

_____

_____

c Is this collision an elastic collision? Explain your answer.

_____

9 Two ball bearings with the same mass collide elastically. Before the collision, ball 1 has velocity $5.5\,\mathrm{m\,s^{-1}}$ and ball 2 has velocity $2.5\,\mathrm{m\,s^{-1}}$ (also in the positive direction). After ball 1 catches up to, and collides with, ball 2, ball 1 has a velocity of $2.5\,\mathrm{m\,s^{-1}}$. What is the velocity of ball 2 after the collision?

_____

_____

_____

_____

_____

10 Two ball bearings with different mass collide elastically. Before the collision, ball 1 has velocity $5.5\,\mathrm{m\,s^{-1}}$ and ball 2 has velocity $2.5\,\mathrm{m\,s^{-1}}$ (also in the positive direction). Ball 1 has twice the mass of ball 2. After ball 1 catches up to, and collides with, ball 2, ball 1 has a velocity of $3.5\,\mathrm{m\,s^{-1}}$. What is the velocity of ball 2 after the collision?

_____

_____

_____

_____

11 In an investigation of one-dimensional elastic collisions, John collides two steel ball bearings in various ways. Ball bearing 1 has mass $m_1 = m$ and ball bearing 2 has mass $m_2 = 1.5m$. Calculate the velocity of each ball bearing after each of these different collisions, assuming elastic collisions.

a Ball bearing 1 is moving at $2.0\,\mathrm{m\,s^{-1}}$ to the right, when it collides with ball bearing 2, which is stationary.

b Ball bearing 1 is moving at $2.0\,\mathrm{m\,s^{-1}}$ to the right when it collides with ball bearing 2, which is also moving to the right but at $1.5\,\mathrm{m\,s^{-1}}$.

c Ball bearing 1 is moving at $2.0\,\mathrm{m\,s^{-1}}$ to the right when it collides with ball bearing 2, which is moving to the left at $1.5\,\mathrm{m\,s^{-1}}$.

12 A neutron collides with a stationary carbon nucleus as shown below. A carbon nucleus has 12 times the mass of a neutron. Calculate the velocity (magnitude and direction) of the carbon nucleus after the collision. On the diagram, draw the vectors representing the velocities of the neutron and the carbon nucleus after the collision.

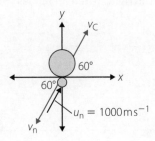

_____

_____

_____

_____

_____

_____

_____

_____

_____

_____

_____

_____

_____

_____

Analysing information from graphical representations of force as a function of time

 SB STUDENT BOOK Pages 166–8

LEARNING GOALS

To use force-time graphs to determine change in momentum of objects

You have already seen in the previous chapter that a graph of applied force as a function of displacement tells you about the energy transferred by the force. This is because $W = \vec{F}s$, so the area under the curve in an F vs s graph is the work done. In this chapter we have seen that force is also related to momentum, via the equation $\Delta \vec{P} = \vec{F}\Delta t$. So the area under an F vs t graph gives the momentum transferred.

1 Jai is pushing a shopping trolley with a force shown on the graph below. Find the momentum transferred by Jai to the trolley.

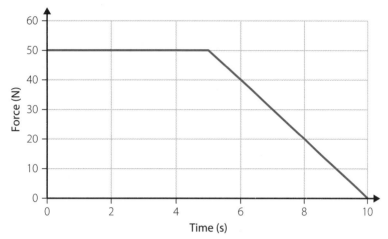

2 The graph below shows the force applied to a ball by a racquet as a function of time. Estimate the total momentum change of the ball due to this force.

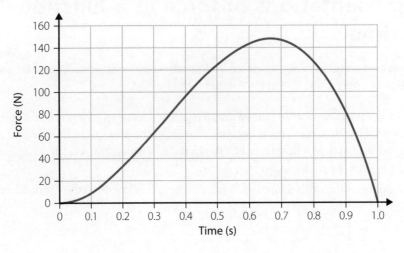

_____

_____

_____

_____

_____

_____

_____

_____

3 The graph below shows the force applied to a toy car as a function of time.

    a Calculate the total momentum change of the car.

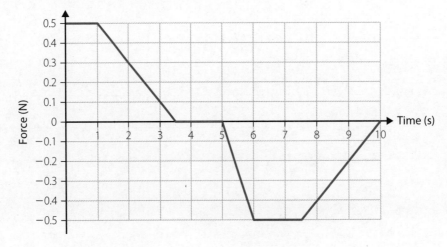

_____

_____

_____

_____

_____
_____
_____
_____

b If the toy car has a mass of 100 g and started at rest initially, what is its final velocity?

_____
_____

4 Sinait is learning to drive, and is practising stopping and starting. Initially the force applied to the car by the road is zero, and it increases steadily over 2 s to a force of 500 N in the positive direction, and then stays at a steady 500 N for 2 s. The force is instantaneously reduced to zero as Sinait takes her foot off the accelerator for 1 s. She then brakes, so that a constant force of 750 N is applied in the negative direction for 2 s, before taking her foot off the brake.

a Draw the force applied to the car as a function of time in the axes below.

b Calculate the total momentum change of the car.

_____
_____
_____
_____
_____

c If the car has a mass of 1200 kg and starts at rest, what is its speed after 2 s?

_____
_____

d What is the car's speed after 5 s?

_____
_____

e What is the car's speed at the end of the time period described?

_____

 **Impulse**

STUDENT BOOK
Pages 166–73

LEARNING GOALS

Understand the relationship between force, time and impulse in collisions

When a force acts on an object it changes both the energy and the momentum of the object. The change in energy is called the work, and the change in momentum is called the impulse.

1 Define 'impulse' and give its units.

_____

_____

2 Explain how impulse can be found from a force vs time graph.

_____

_____

3 What does the gradient of a graph of an object's momentum as a function of time represent?

_____

_____

4 David wants to increase the distance he can hit the ball in golf. Describe two ways he can do this by increasing the impulse given to the ball.

_____

_____

5 Su-Hyun hits a golf ball with an average force of 30 N for 0.1 s.

a What impulse is transferred to the ball?

_____

b If Su-Hyun applied the same average force, and transferred an impulse of $10 \, \text{kg m s}^{-1}$ to the ball, for how long would she have to apply the force?

_____

c If Su-Hyun wished to transfer an impulse of $10 \, \text{kg m s}^{-1}$ but in only 0.1 s, what average force would she need to apply?

_____

6 It takes 12 s for a car with mass 1900 kg to reach $100 \, \text{km h}^{-1}$.

a What impulse does the car receive?

_____

_____

b What average force does the car experience during this time?

_____

_____

c When the same car travelling at $100 \, \text{km h}^{-1}$ collides with a tree, it takes only 0.75 s to come to a complete stop. What impulse does the car receive in this collision?

_____

_____

9780170449595

d What average force does the car experience during this time?

_____

_____

e What average force does a passenger (mass 65 kg) experience during this collision, assuming the passenger takes the same time to come to a stop as the car?

_____

_____

_____

f Explain, in terms of force and impulse, how safety features such as seatbelts and crumple zones protect the people in a car during a collision.

_____

_____

_____

_____

7 The graph below shows the force as a function of time acting on an object during a collision. On the set of axes provided, plot the change in momentum of the object as a function of time $\Delta\vec{P}$ vs Δt. This is the same as the momentum the object would have as a function of time, assuming it started from rest.

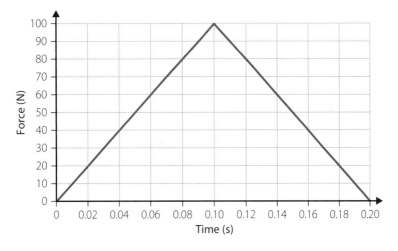

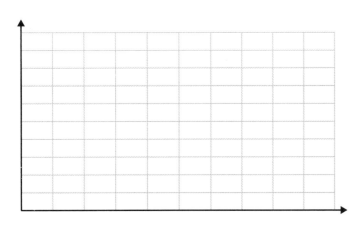

_____

_____

_____

_____

_____

_____

_____

8 Impulse is a vector quantity. For each of the following situations, state the direction of the impulse in each collision, and draw a vector diagram showing the initial momentum, final momentum and the impulse to explain your answer. Show the vector subtraction, $\Delta \vec{p} = \vec{p}_{final} - \vec{p}_{initial}$, used to find the impulse.

   a A stationary train carriage is shunted from behind by a locomotive. Take the direction of travel of the carriage after the collision to be positive.

   b A tennis ball is initially travelling in the positive x direction, when it is hit with a racquet and rebounds in the negative x direction.

   c An ice hockey puck is sliding along the ice to the north, when it is struck. It then slides off to the north-west.

LEARNING GOALS

Analyse momentum and energy in both elastic and inelastic collision situations

In any interaction, and hence in any collision, momentum is conserved. But the kinetic energy of the two colliding objects is rarely conserved – it can change form, for example into stored elastic energy and into thermal energy. If you strike a piece of wood repeatedly with a hammer you will find that the wood gets warm. This is kinetic energy being converted into thermal energy due to the collisions with the hammer.

1 Explain the difference between an elastic collision and an inelastic collision. What is meant by a perfectly elastic collision and a perfectly inelastic collision?

2 Give an example of a perfectly elastic collision.

3 Give an example of a perfectly inelastic collision.

4 Ellyse catches a fast-moving cricket ball.

a Is this an elastic or inelastic collision? Justify your answer.

b The ball has a mass of 150 g and is moving at $15\,\mathrm{m\,s^{-1}}$ when Ellyse, mass 65 kg, catches it. If she is stationary, and if no other forces were acting, what would be her speed after catching the ball?

c Why is Ellyse's actual speed after the collision much lower than this?

5 Two ball bearings with the same mass collide. Before the collision, ball 1 has velocity $6.5\,\text{m}\,\text{s}^{-1}$ and ball 2 has velocity $4.5\,\text{m}\,\text{s}^{-1}$ (also in the positive direction). After ball 1 catches up to, and collides with, ball 2, ball 1 has a velocity of $5.0\,\text{m}\,\text{s}^{-1}$.

   a What is the velocity of ball 2 after the collision?

_____

_____

_____

_____

_____

   b Is this an elastic or inelastic collision? Justify your answer.

_____

_____

_____

_____

_____

6 Jackson drops a 0.5 kg lump of clay from a height of 1.3 m on to the floor, where it sticks.

   a What type of collision is this?

_____

   b Ignoring air resistance, calculate the speed with which the clay hits the floor.

_____

_____

_____

   c What is the momentum change of the clay as a result of the collision?

_____

   d What is the momentum change of Earth as a result of the collision?

_____

_____

   e If we take Earth as stationary before the collision, what is its speed after the collision? The mass of Earth is $6.0 \times 10^{24}\,\text{kg}$.

_____

_____

7 Nuclear decay can be modelled as a perfectly inelastic collision in reverse. A carbon-14 nucleus with a mass of $2.3 \times 10^{-26}\,\text{kg}$ decays by ejecting an electron. This converts carbon-14 into nitrogen-14. The ejected electron has a kinetic energy of $2.5 \times 10^{-14}\,\text{J}$. Treat the carbon-14 nucleus as initially stationary.

   a What is the speed of the ejected electron? ($m_\text{e} = 9.1 \times 10^{-31}\,\text{kg}$).

_____

_____

b What is the momentum of the ejected electron?

_____

c With what speed does the nitrogen nucleus recoil?

_____

_____

_____

_____

8 A 3500 kg weather satellite is moving at $3.1\,\text{km}\,\text{s}^{-1}$ when it is struck by a 250 kg meteorite moving at $15\,\text{km}\,\text{s}^{-1}$. The velocity of the meteorite is perpendicular to that of the satellite.

Assuming the collision is perfectly inelastic, calculate the velocity of the satellite after the collision. Find the angle between the final velocity and the initial velocity of the satellite.

_____

_____

_____

_____

_____

_____

_____

_____

Module two: Checking understanding

Circle the correct answer for questions 1–5.

1 If two equal forces are applied to an object – one from the south and one from the west – then the total force acting on the object as a result of these two forces:

   A must be directed north-east.

   B cannot be determined without knowing the magnitude of the forces.

   C must be directed south-west.

   D will be zero because they will cancel each other out.

2 If a tension force of T is applied along a rope at $10°$ to the vertical, then the vertical component of T will be given by the expression:

   A $T\sin 10°$

   B $T\cos 10°$

   C $T\tan 10°$

   D T

3 A small magnet is suspended vertically from a piece of string. When a second magnet is brought to a position below and to the side, the string of the suspended magnet now makes an angle with the vertical. Which statement correctly lists the contact and non-contact forces acting on the suspended magnet?

   A Tension (contact force) and magnetic force (non-contact force)

   B Tension (contact force) and gravitational force (non-contact force)

   C Tension (non-contact force), magnetic force (non-contact force) and gravitational force (contact force)

   D Tension (contact force), magnetic force (non-contact force) and gravitational force (non-contact force)

4 A large portrait at the Art Gallery of New South Wales is hung from two wires connected at the corners, both making angles of $20°$ with the vertical. Which of the following statements is true of the forces acting on the portrait?

   A The net force is zero but the tension force in each of the two wires is less than the weight force.

   B The total upwards force due to the tension in the two wires is less than the weight force.

   C The net force is zero and the tension force in each of the two wires is equal to the weight force.

   D The total upwards force due to the tension in the two wires is greater than the weight force.

5 A Physics teacher jumps from a bench in the Science laboratory. Which statement correctly describes the forces as he falls?

   A The force that Earth exerts on the teacher is greater in magnitude than the force the teacher exerts on Earth, which is why the teacher accelerates more than Earth.

   B The force that Earth exerts on the teacher is unrelated to the force the teacher exerts on Earth because it depends on the relative mass of each object.

   C The force that Earth exerts on the teacher is equal in magnitude to the force the teacher exerts on Earth but the teacher accelerates more because his mass is less.

   D The force that Earth exerts on the teacher is equal in magnitude to the force the teacher exerts on Earth, which is why both objects experience equal acceleration.

6 An elevator engine is able to lift the elevator, of mass $480\,kg$, and 12 passengers with an average mass of $65\,kg$, at a rate of $0.8\,m\,s^{-1}$ when operating at capacity.

   a Calculate how much work would be done by the engine raising this load $40\,m$.

   _____

   _____

   b What would be the increase in gravitational energy of an average passenger during this journey?

   _____

   _____

c Determine the output power of the engine when operating at capacity.

_____

_____

7 On the International Space Station an astronaut is using two tennis balls, one covered with Velcro, to model collisions. The ball without Velcro (ball A) has a mass of 40 g and the ball with Velcro (ball B) has a mass of 60 g. In one demonstration she pushes ball A to the left at $0.5\,\mathrm{m\,s^{-1}}$ and ball B to the right at $0.75\,\mathrm{m\,s^{-1}}$ so that they collide.

a Assuming that the two balls stick together, calculate their velocity after the collision.

_____

_____

_____

_____

b Calculate the impulse of ball A.

_____

_____

c Show that this collision is inelastic.

_____

_____

_____

_____

d A slow motion film of the collision reveals that the collision lasted 0.2 s. What force did ball B exert on ball A? How does this force compare to the force ball A exerted on ball B?

_____

_____

_____

8 A 5 kg wooden block slides down a metal ramp, set at an angle of 8°, at $0.2\,\mathrm{m\,s^{-1}}$. Vector arrows representing forces related to the block's motion are shown in the diagram below.

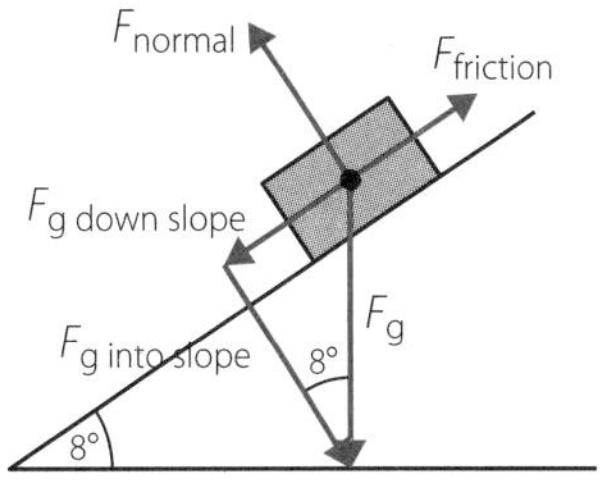

a In the diagram you will note that the arrow labelled $F_{g\,\text{down slope}}$ is equal in length to the arrow labelled F_{friction} and the arrow labelled $F_{g\,\text{into slope}}$ is equal to the one labelled F_{normal}. Explain why this must be the case in this example.

_____

_____

_____

b Using trigonometry, write an expression for $F_{g\text{ down slope}}$ in terms of the mass of the block and the angle of the slope.

_____

_____

_____

c Using trigonometry, write an expression for $F_{g\text{ into slope}}$ in terms of the mass of the block and the angle of the slope.

_____

_____

_____

d Using trigonometry and the equation $F_{friction} = \mu \times F_{normal}$, write an expression for $F_{friction}$ in terms of μ, the mass of the block and the angle of the slope.

_____

_____

_____

e Use your answers to **b**, **c** and **d** to find a value for μ in this situation.

_____

_____

_____

_____

_____

Reviewing prior knowledge

1 What is the definition of a wave?

_____

2 Match each term with the appropriate description by writing the number of the description next to the term.

A	Conduction		1	Heat coming from the Sun
B	Convection		2	A metal pan heating up on a stove
C	Radiation		3	Warm air circulating in a room

3 In your own words, describe the following parts of a wave.

  a Wavelength

_____

  b Frequency

_____

  c Period

_____

4 State the type of wave based on the description of its movement.

  a The particles oscillate perpendicular to the direction of energy transfer along the wave.

_____

  b The particles oscillate along the same direction as the direction of energy transfer in the wave.

_____

5 A pencil in a glass of water appears as if it is broken into two parts. This is an example of what?

_____

6 The Coney Island Funhouse at Luna Park has a number of different shaped mirrors. When a person walking through sees themselves in the mirrors, this is an example of what?

_____

7 What are some uses of electromagnetic waves in today's world? Identify three electromagnetic waves and where on the spectrum they would be found.

_____

_____

_____

_____

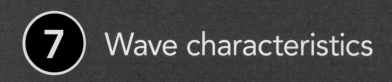
 Investigating waves

STUDENT BOOK
Pages 186–97

How do you hear birds sing? How do you see stars? It is due to the transfer of energy. In the case of hearing a bird sing, it is the transfer of energy through a medium with which the energy can interact as it passes through. No such medium is required to see the stars. In this worksheet you will undertake two experiments to investigate the role the type of medium has in this energy transfer.

INVESTIGATION

Role of the medium in the propagation of mechanical waves

AIM

To investigate the role of the medium in the propagation of the wave

HYPOTHESIS

1 Read the method to form an idea of how the experiment will work. Devise a suitable hypothesis that will demonstrate the different effects the medium will have on the propagation of the wave.

_____

_____

RISK ASSESSMENT

2 Complete a risk assessment that covers at least three risks that are specifically associated with this investigation.

RISK ASSESSMENT

What are the risks in doing this experiment?	How can you manage these risks to stay safe?

MATERIALS

- Oven rack
- String
- Scissors
- Metal spoon
- Wooden spoon

METHOD

1 Tie a 1 m length piece of string to each side of the oven rack so that the string is parallel with the bars of the oven rack. Ensure the slack in the string is evenly distributed at each end.

2 Wrap the slack string at each end to a finger on each hand.

3 Place your fingers with the string wrapped around them in your ears.

4 Have another person hit the rack with one of the spoons.

5 Record what you hear.

6 Repeat the experiment with the other spoon.

3 Devise a suitable method for observing mechanical waves through the medium of the air rather than the string. In your method, make sure that you identify the independent, dependent and any control variables that you are looking at. Think about the different ways that the energy is transferred: can you make it go in one direction or is it going to spread out?

RESULTS

4 Record the results of your experiment in the space below.

5 Consider the following questions and discuss them with your group. Use the space below to make notes.

   a What effect did the different medium have on the transfer of energy in your experiment?

   b Were you able to vary the direction of travel in the different mediums?

   c If you were going to carry out this investigation again, what changes would you make to ensure a more valid/ reliable investigation?

_____

_____

_____

_____

_____

_____

_____

_____

_____

_____

CONCLUSION

6 Did your results support your hypothesis? Write a suitable conclusion to your investigation, making sure that you are relating it back to your hypothesis.

_____

_____

_____

INVESTIGATION

Transfer of energy by a mechanical wave

AIM

To observe how energy can be transferred by a mechanical wave

HYPOTHESIS

7 Read the method to form an idea of how the experiment will work. Devise a suitable hypothesis for this experiment.

_____

_____

8 Complete a risk assessment that covers at least three risks that are specifically associated with this investigation.

What are the risks in doing this experiment?	How can you manage these risks to stay safe?

MATERIALS

• Masking tape   • Wooden skewers   • Jelly babies   • Video recording equipment such as a mobile phone

METHOD

1 Roll a piece of masking tape flat on the floor or a counter for 3 m, sticky side up.

2 Position 15 wooden skewers at 15 cm intervals along the masking tape, with an equal length of each skewer extending beyond each side of the tape. Leave at least 20 cm of tape at one end with no skewers on it.

3 Place another piece of masking tape over the top of the skewers to fix them in place.

4 Add a jelly baby to each end of each skewer, making sure that they are evenly weighted on both sides. This can be done by sliding one of the jelly babies along the skewer.

5 Fix one end of the tape to a solid object such as a chair or table, or hang it from a suitable height such as the ceiling.

6 Using the skewer at the free end of the tape, impart a twist to the tape. Observe and record what happens.

7 Now move the skewer at the free end back and forth instead.

RESULTS

9 Record the results of your experiment in the space below.

10 Consider the following questions and discuss them with your group. Use the space below to make notes.

   a What happened when you twisted the skewer?

   b Was what happened independent of the direction it was twisted?

   c Both of the waves you created with this system were transverse waves. Explain how these are different from longitudinal waves.

_____

_____

_____

_____

_____

_____

CONCLUSION

11 Did your results support your hypothesis? Write a suitable conclusion to your investigation, making sure that you are relating it back to your hypothesis.

_____

_____

_____

INVESTIGATION

Differences between transverse and longitudinal waves

AIM

To demonstrate the differences between transverse and longitudinal waves

12 Construct a suitable research question for this investigation

_____

_____

RISK ASSESSMENT

13 Complete a risk assessment that covers at least three risks that are specifically associated with this investigation.

RISK ASSESSMENT

What are the risks in doing this experiment?	How can you manage these risks to stay safe?

• Slinky spring • Ribbon • Video recording device, such as mobile phone

METHOD

1 Tie pieces of ribbon onto the slinky at 30 cm intervals, making sure that the ribbons can be seen on the same side.
2 Move two people in your group so they are standing 3–4 m apart, each holding an end of the slinky.
3 Set up the recording device so you can clearly see the spring and the bits of ribbon.
4 One person rapidly moves their end of the slinky up and down to send a wave through the slinky.
5 Each person then pulls back on the slinky and then releases the stretch to send another wave down the spring.

RESULTS

14 Watch the video recording to analyse the position of the ribbons as each type of wave travels down the slinky. Make a note of the movement of the wave in relation to the movement of the pieces of ribbon.

_____

_____

_____

_____

_____

_____

_____

_____

_____

_____

DISCUSSION

15 Consider the following questions and discuss them with your group. Use the space below to make notes.

a What happened to the ribbon? Did it move with the wave or oscillate back and forth around the same spot?
b How did the movement of the ribbon differ for the two types of wave?
c How has this shown that you can transfer energy without transferring mass?

_____

_____

_____

_____

_____

_____

_____

_____

_____

_____

Differences between mechanical and electromagnetic waves

16 We have discussed and shown that mechanical waves require a medium to travel through. This is the main difference between mechanical and electromagnetic waves. Why then does an electromagnetic (EM) wave not require a medium to travel through?

_____

_____

_____

17 In the movie _Alien_, the tagline is 'In space no one can hear you scream'. Analyse this statement.

_____

_____

_____

_____

_____

_____

_____

_____

_____

18 Research the parts of the EM spectrum and list your results in the table below. Include information such as wavelength, frequency and uses.

Name of wave	λ	Frequency	Uses

| WS | 7.2 | Graphs of displacement |

LEARNING GOALS

- Interpret graphs of waves and their properties
- Draw graphs of wave behaviour
- Distinguish between graphs of displacement versus time and displacement versus distance

When discussing waves, it is important to understand the terminology that is used and how to interpret the data that is given. These include the displacement, amplitude, frequency, wavelength, period and velocity of the wave. By representing waves graphically, we can clearly show how the different characteristics of waves are related. As waves are described both by how they change in time and how they change in space, we use two types of graphs: graphs of displacement as a function of time, and graphs of displacement as a function of position.

Displacement versus time

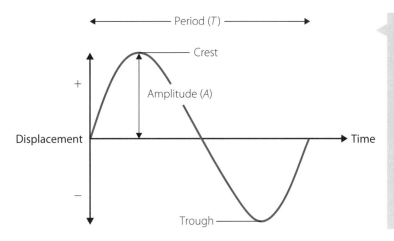

A displacement versus time graph of a mechanical wave represents the displacement of one particle of the medium as it experiences a wave disturbance over time.

1 Draw a displacement time graph that illustrates 2 complete oscillations of a wave with an amplitude of 30 cm and a period of 0.25 s.

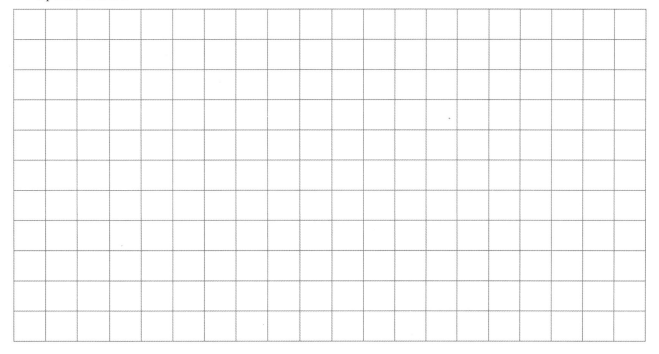

2 The following trace was taken from a cathode-ray oscilloscope (CRO) with the sweep time set to 4 ms per division. Find the period of the wave.

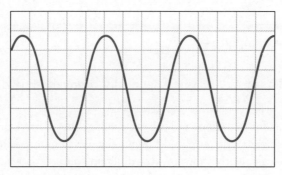

---

3 The following graphs all show the displacement versus time for four different waves all with the wavelength 0.75 m. For each wave determine the period, frequency and amplitude.

a

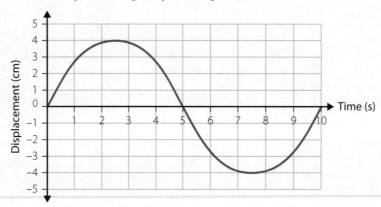

   i Period _____   **ii** Frequency _____   **iii** Amplitude _____

b

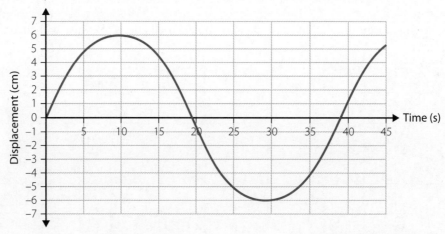

   i Period _____   **ii** Frequency _____   **iii** Amplitude _____

c

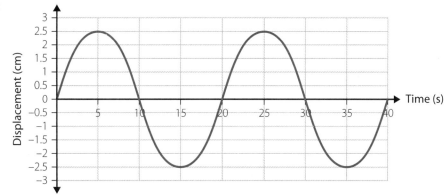

   i Period _____   **ii** Frequency _____   **iii** Amplitude _____

d

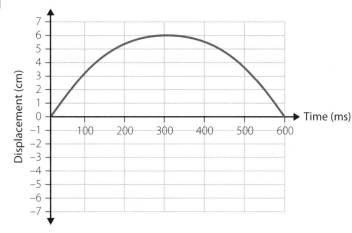

   i Period _____   **ii** Frequency _____   **iii** Amplitude _____

Displacement versus distance

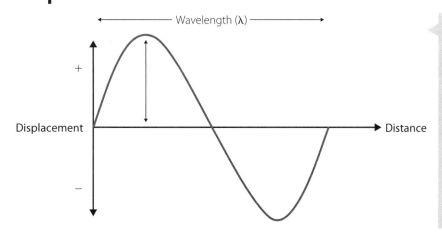

The displacement versus distance graph of a mechanical wave shows the displacement of all the particles of the medium at an instant in time.

4 A wave travelling at $36\,\mathrm{m\,s^{-1}}$ has a wavelength of $13\,\mathrm{m}$ and an amplitude of $6\,\mathrm{m}$. Choose appropriate axes to draw the displacement versus distance graph for this wave.

5 The following graphs represent a sound wave as displacement of particles against distance, the variation of air pressure against distance, and the position of the particles as the wave passes through the area.

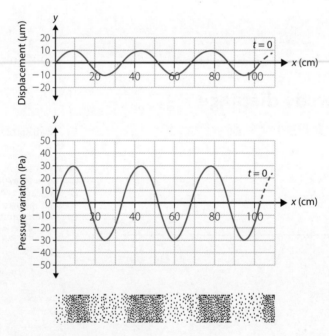

From the information above, find the following.

a The wavelength _____

b The maximum pressure caused by the wave _____

c Is this a longitudinal or transverse wave? Explain your answer.

_____

_____

_____

_____

Be able to describe a wave quantitatively

Apply the formulas for working out the velocity and period of a wave

We have looked at the features of waves in the preceding worksheets, but how do we link them together mathematically? It is important to be able to describe the characteristics of a wave to see how they relate to each other.

For this worksheet we will be looking at the formulas $f = \dfrac{1}{T}$ and $v = f\lambda$.

HINT

When working out answers, it is important to remember to follow the steps in solving a physics problem and show *all* of working out – and remember to include units on all quantities.

1 Eric plucks a guitar string, making it vibrate. It takes 0.05 s for the string to complete one full oscillation. What is the frequency of the wave?

_____

2 The same guitar string is now oscillating at a frequency of 100 Hz. What is the period of these oscillations?

_____

_____

3 The equation to find the period from the frequency is referred to as an inverse relationship. Explain why this is the case.

_____

_____

_____

_____

4 If a wave has a period of 3 s, and a wavelength of 1.4 m, how fast will this wave travel?

_____

_____

_____

_____

5 Complete the table below for the electromagnetic spectrum. The speed of light (electromagnetic waves) is $3.0 \times 10^8 \, \text{m s}^{-1}$.

Type of wave	λ	Frequency
Radio	13 m	
		3.25×10^{10} Hz
Red light	680 nm	
Blue light		660 THz
	0.350 pm	
Gamma ray		4.6×10^{19} Hz

6 **a** You are intending to go out for a walk when you see dark clouds come in and then lightning. You still want to go out, so you decide to see if you can find out how far away the lightning is from you. You count the time it takes for the sound to reach you after the flash, and reach 3 s. If the speed of sound in air is $340 \, \mathrm{m\,s^{-1}}$, how far away is the lightning?

_____

_____

b Taking the speed of light to be $3 \times 10^8 \, \mathrm{m\,s^{-1}}$, calculate how much time it took for the light to reach you from the flash.

_____

_____

7 You are sitting at the end of the pier, fishing. While waiting for the fish to bite you see nine waves going past per minute and measure the distance between the wave crests to be 15 m. Find:

  a the period of the waves

_____

  b the frequency of the waves

_____

  c the velocity of the waves.

_____

8 **a** A sound wave has a period of 7 ms in air (the speed of sound in air is given as $340 \, \mathrm{m\,s^{-1}}$). What is its wavelength in air and in a copper rod, in which sound has a speed of $3750 \, \mathrm{m\,s^{-1}}$?

_____

_____

_____

_____

  b Explain why the wavelength and speed change, but not the frequency.

_____

_____

_____

_____

_____

9 If a radio station has a frequency of 101.7 MHz, what would be the wavelength of these waves?

_____

_____

8 Wave behaviour

INQUIRY QUESTION: HOW DO WAVES BEHAVE?

WS 8.1 Investigating phenomena affecting the behaviour of waves

STUDENT BOOK
Pages 209–24

LEARNING GOALS

Identify the different wave behaviours

Predict the outcome of wave behaviour in a number of different situations

When you make a call on your mobile phone, do you think about how the information gets to you? When you are using wi-fi does your location affect the signal? How do we know what happened in the seconds after the big bang? It is the understanding of the behaviour of waves that allows us to live in such a technologically advanced society. The wave phenomena of reflection, refraction, diffraction and superposition will be explored in this worksheet.

Reflection

1 Write a definition of the law of reflection.

_____

_____

2 Fill in the spaces in the diagram below.

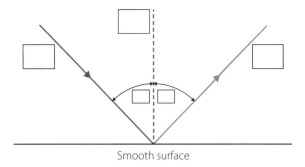

Smooth surface

Reflection off a concave surface

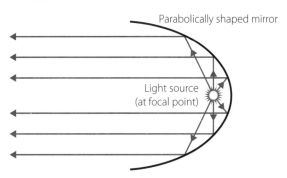

Parabolically shaped mirror

Light source
(at focal point)

The diagram above shows the application of reflections on concave surfaces, in this case a car headlight.

3 Identify two other applications for reflections off a concave surface. Explain.

_____

_____

_____

Reflection off a convex surface

4 What is the difference between the image formed by a convex mirror and the image formed by a concave mirror? Provide an example of a use for a concave mirror.

_____

_____

5 Draw a diagram of a convex mirror in the space below, showing the lines of reflection. What would this do to the image reflected?

_____

_____

INVESTIGATION

Reflection of light

AIM

To investigate the reflection of light waves in curved and flat mirrors at various distances

MATERIALS

• 1 large shiny metal spoon   • 1 large flat metal spatula or butter knife

RISK ASSESSMENT

6 Complete a risk assessment table for three risks associated with this experiment.

RISK ASSESSMENT

What are the risks in doing this investigation?	How can you manage these risks to stay safe?

METHOD

1 Hold the spoon as far away from you as you can. Make sure that the concave side is towards you. Make a note of the reflection in the spoon. Is it upside down or the right way up? Is it magnified or not?

2 Bring the spoon towards you and note any changes in the reflection.

3 Turn the spoon the other way around so the convex side is facing you and repeat steps 1–2.

4 Using the flat surface of the spatula/knife, repeat steps 1–2.

RESULTS

7 Complete a suitable table with your results.

DISCUSSION

8 In which image would you have found the rays diverging, converging or parallel?

9 Draw a diagram and relate the shape of the surface to the behaviour of the rays.

CONCLUSION

10 Relate your results to the aim.

Refraction

11 In your own words explain why the wave front changes direction when a wave hits the surface of a new medium.

12 Complete the diagrams below showing the direction of the wave.

a Warm air (1)

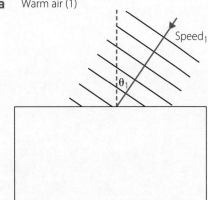

b Warm air (1)

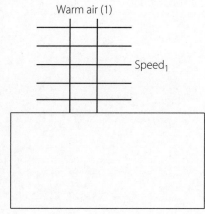

13 Give some examples of where you would see refraction of light waves in everyday situations. For each example, explain what is happening.

_____

_____

_____

_____

Diffraction

When a wave passes through an opening, such as a door or window, it will spread out into the space behind. This is called diffraction.

14 Fill in the spaces in the following sentences.

High pitched sounds have _____ *wavelengths. These are diffracted*

_____ *than lower pitched sounds with* _____ *wavelengths.*

15 Complete the drawings below showing waves with different wavelengths diffracting around various objects.

a
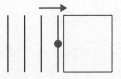

b
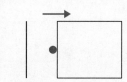

c

d
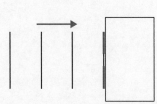

e
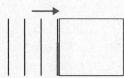

f

Wave superposition

16 Define 'wave superposition'.

_____

_____

_____

_____

17 Define 'constructive interference' and 'destructive interference'.

_____

_____

_____

_____

_____

_____

18 Provide examples of constructive and destructive interference.

_____

_____

_____

_____

_____

_____

19 Draw the following waves to the same scale on the grid below. Choose a suitable scale and start the waves at maximum amplitude.

a A wave with a wavelength (λ) of 40 mm and an amplitude of 20 mm

b A wave with a λ of 40 mm and an amplitude of 30 mm

c The resultant wave

20 Choose a suitable scale and draw the following waves on the grid below.

   a A wave with a λ of 50 mm and an amplitude of 20 mm. Start at maximum amplitude

   b A wave with the same wavelength and amplitude, out of phase to the wave from part **a**

   c The resulting wave

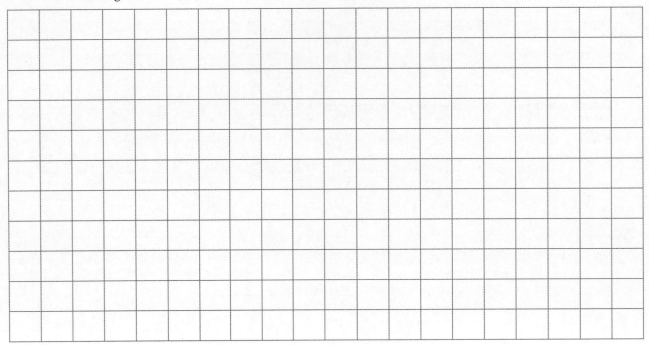

21 On the graph below, draw in the resulting wave from the two waves shown.

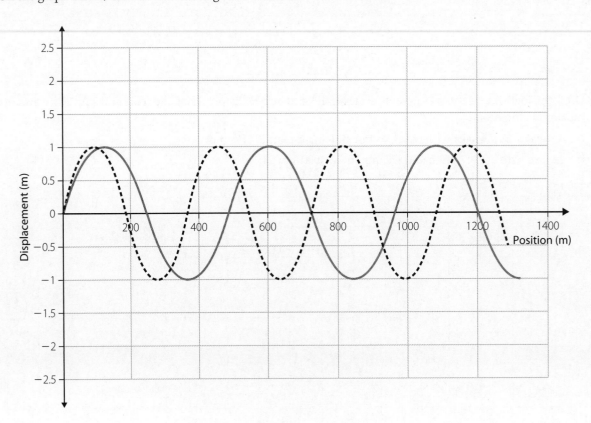

Distinguishing between progressive and standing waves

LEARNING GOAL

Identify what is a standing wave and a progressive wave and the difference between them

1 In your own words, compare a standing wave and a progressive wave.

_____

_____

_____

_____

2 Complete the table for the differences in progressive and standing waves.

Property	Standing wave	Progressive wave
Frequency		
Amplitude of oscillations		
Amplitude of wave		

INVESTIGATION

Showing the difference between standing waves and progressive waves

AIM

To use a rubber tube to show the differences between standing waves and progressive waves

MATERIALS

- Rubber tube
- Mobile phone camera

METHOD

1 Stretch the tube over a distance of at least 2 m.
2 Hold one end of the tube still.
3 Send a pulse of a wave down the tube by raising the free end up and bringing it down quickly. Make a video recording of the resulting wave on your mobile phone.
4 Send multiple pulses down the tube by repeatedly moving the free end up and down in a smooth motion.
5 Continue to send waves down the tube until it appears that the wave is no longer moving. You might need to adjust the frequency to do this. Make a video recording of the wave.

3 Analyse the two different video recordings of the different waves, make a note of what appears to be happening to the rubber tube.

_____

_____

_____

_____

_____

DISCUSSION

4 What was the difference between the progressive wave and the standing wave?

_____

_____

_____

_____

_____

5 If you change the tension in the tube but do not change anything else, does this make a difference to the frequency of the standing wave?

_____

_____

_____

_____

_____

CONCLUSION

6 Have you been able to show the differences between progressive and standing waves?

_____

_____

_____

_____

Researching resonance in mechanical systems

LEARNING GOALS

Describe resonance in a system

Identify a real-world example of resonance

1 Describe 'resonance' in your own words.

_____

_____

INVESTIGATION

Secondary-sourced investigation into resonance

AIM

To investigate a real-world example of resonance using valid and reliable sources

2 What is your example?

_____

_____

3 How does it demonstrate resonance?

_____

_____

_____

_____

_____

_____

_____

_____

_____

_____

4 Keep track of the sources you used by recreating the table below in a Word or Excel document.

Title	Type of source, (for example journal article)	Citation (how this would be cited in a reference list)	Why is it a valid source?

INQUIRY QUESTION: WHAT EVIDENCE SUGGESTS THAT SOUND IS A MECHANICAL WAVE?

STUDENT BOOK
Pages 235–9

WS 9.1 Investigating pitch and loudness

LEARNING GOALS

Identify the difference between pitch and loudness

Describe the relationship of pitch and loudness to wavelength and amplitude of a wave

Sound is a mechanical wave that requires a medium to travel through. It can behave like any other mechanical wave – it can be reflected, refracted or diffracted. The pitch (low to high) and volume (quiet to loud) can be displayed using an oscilloscope, which shows the movement of the particles of the sound wave in the form of a transverse wave or sinusoidal curve. An example of a soundwave as shown by an oscilloscope is below.

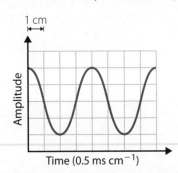

INVESTIGATION

Investigating pitch and volume using an oscilloscope

AIM

To relate the pitch and volume of a sound wave to its characteristics, as represented by an oscilloscope

HYPOTHESIS

1 Write a suitable hypothesis for this investigation.

_____

_____

_____

RISK ASSESSMENT

2 Complete the risk assessment below.

What are the risks in doing this investigation?	How can you manage these risks to stay safe?

MATERIALS

• Signal tone generator (these can be downloaded onto a smart phone with apps such as Phyphox or Google Science Journal)   • Oscilloscope (can be found either through the apps mentioned above or through free online sites)   • Tuning forks of different frequencies   • An extra person, to whistle

METHOD

3 In the space below write a suitable method for this investigation.

_____

_____

_____

_____

_____

_____

RESULTS

4 Record your results.

_____

_____

_____

_____

DISCUSSION

5 Discuss your results.

_____

_____

_____

_____

Modelling sound

Show that sound is a longitudinal wave

Identify the areas of compression and rarefraction in a longitudinal wave

1 When sound travels through a medium the particles oscillate in a longitudinal direction; that is, parallel to the direction of energy transfer. Below is a diagram of the areas of low and high pressure of a wave travelling through a medium. Complete the table below to identify:

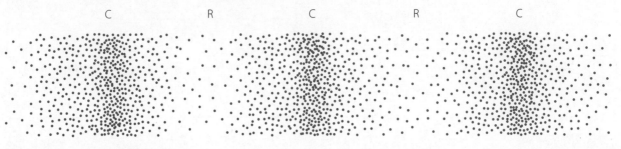

C   R   C   R   C

a what the letters C and R stand for

b which letter refers to areas of low pressure and which to areas of high pressure.

	C	R
Meaning of letter		
Pressure level		

INVESTIGATION

Modelling sound as a longitudinal wave

AIM

To model sound as a longitudinal wave using video analysis

METHOD

2 In the space below, write a suitable method for this investigation. Think about what you could use to demonstrate this clearly. Would a rubber band work or is a slinky spring clearer?

_____

_____

_____

_____

_____

_____

_____

3 In your video, what do you see happening with the different particles (parts of the wave)?

_____

_____

_____

_____

DISCUSSION

4 How would you relate this to the fact that sound is a mechanical wave?

_____

_____

_____

_____

_____

_____

5 How would this affect the pressure within the system that the sound is travelling through?

_____

_____

_____

_____

_____

_____

Calculating the intensity of sound over distance

STUDENT BOOK
Pages 242–5

LEARNING GOALS

Understand that sound follows the inverse square law

Apply the correct formula to show how sound changes intensity over distance

1 As with all waves, the energy carried by a sound wave dissipates over its surrounding area. If there is less energy in a particular area, the intensity of the sound drops. Explain how this happens.

_____

_____

_____

_____

2 A teacher is talking in the classroom. The students at the front of the class are sitting 2.00 m away and can hear the teacher talking at an intensity of $3.00 \times 10^{-7}\,\mathrm{W\,m^{-2}}$. What intensity of sound is heard by the students 5.00 m away at the back of the class?

_____

_____

_____

_____

_____

3 Leo is watching a formula one race. He is standing at the front, against the safety barrier 15 m away from the track, but his ears are starting to hurt because the intensity of the cars is $3.20\,\mathrm{W\,m^{-2}}$. He wants to move back to reduce the intensity of the sound.

a How far back does Leo have to move so the intensity level is similar to a busy road – $8.00 \times 10^{-4}\,\mathrm{W\,m^{-2}}$?

_____

_____

_____

_____

_____

b What distance away would Leo need to be at to hear 15% of the intensity of the original sound?

_____

_____

_____

_____

4 Rebecca observes a jackhammer at a building site. Rebecca is 60 m away and hears the intensity at $2.00 \times 10^{-3}\,\mathrm{W\,m^{-2}}$. At what intensity does the operator of the jackhammer hear the sound? If the operator was not wearing any hearing protection, what would happen? Explain.

_____

_____

_____

_____

_____

_____

_____

WS 9.4 Investigating reflection of sound waves

STUDENT BOOK
Page 246

LEARNING GOALS

Identify the conditions required for sound to reflect

Understand that sound will reflect according to the law of reflection

1 When we have a reflected soundwave, we have an echo. Name a use for sound reflection and explain how it works.

_____

_____

_____

INVESTIGATION

Observing the reflection of sound waves

AIM

To demonstrate that waves can reflect and be amplified by the superposition of the waves

MATERIALS

• Rubber bands   • Empty container such as an ice-cream tub   • Cloths

METHOD

1 Stretch the rubber band between your fingers and pluck it. Record your observations of the sound.

2 Stretch some rubber bands across the open top of your container and pluck them. Record your observations of the sound.

3 Place some of the cloths into the container and pluck the rubber bands. Record your observations of the sound.

RESULTS

2 Record your observations and results in a table.

Equipment used	Observations
Rubber band across your fingers	
Rubber band across an ice-cream container	
Rubber bands across an ice-cream container with cloth in it	

3 Why did the sound of the rubber band differ in each situation?

_____

_____

_____

_____

_____

_____

_____

CONCLUSION

4 What can you conclude from your results?

_____

_____

_____

_____

_____

_____

Investigating the behaviour of standing waves on strings

STUDENT BOOK
Pages 249–59

1 How do you create a note in a pipe or a string?

_____

_____

_____

2 What are the areas of maximum movement and the areas of minimum/no movement along the string or in the pipe called, and what causes them?

_____

_____

_____

3 Complete the table of the first four harmonics on a string fixed at both ends.

Vibration mode	Wave pattern	f and λ (in terms of v and l)

4   Complete the table for the first four harmonics for a pipe open at both ends. Show the pressure variation through the pipe as well.

Vibration mode	Particle displacement	Pressure variation	f and λ (in terms of v and l)

5   An experiment is devised to use standing waves to measure mass per unit length of a piece of string. The equipment is set up below.

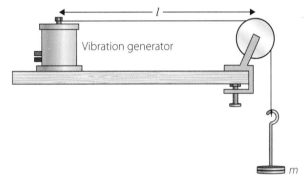

In this experiment, several measured quantities have uncertainties.

a   How would you determine uncertainty in the measured masses that are used in the results table?

_____

b   What would be an appropriate way to determine the uncertainty for the wavelengths measured with a normal ruler?

_____

6   The students used the following method to gather their results.

a   What is the total mass of the mass carrier when fully laden?

_____

b Identify the independent variable and the dependent variable.

____

____

METHOD

1 Set up equipment as shown in the diagram, with the mass carrier empty.
2 Starting with a 50 g mass measure the wavelength.
3 Now increase the mass on the carrier in 50 g steps until there is 250 g mass on the carrier.
4 Record all data in your table.
5 Plot the results onto a suitable graph.

RESULTS

7 The results the students achieved are below. Fill in the gaps using the information in the cells that have already been completed.

Total mass on the end of the string (g)	Tension force (N)	$\frac{\lambda}{2}$ (cm)	λ (cm)	$\frac{3\lambda}{2}$ (cm)	Average λ (cm)
50 ± 1.25	$0.49 \pm 0.012\,25$		30	45	
100 ± 2.5		21			42
150 ± 3.75	$1.47 \pm 0.036\,75$		50.5	73.75	
		28.5	57		57
250 ± 6.25	$2.45 \pm 0.061\,25$	31.75		95.25	63.5
	2.94 ± 0.0735	34.75	69.5		69.5

8 What is the % uncertainty for the mass and tension force? ______________________

9 If we measure the speed of a wave using $v = f\lambda$ and the speed of a wave in a string is given as $v = \sqrt{\frac{T}{\mu}}$ where v = velocity in the string, T = the tension force, which is given by the weight formula ($T = mg$), and μ = the mass per unit length of the piece of string (given by $\frac{\text{mass}}{\text{length}}$ of the piece of string). Expanding the equation, we have $f\lambda = \sqrt{\frac{mg}{\mu}}$, solving for $\lambda^2 = \frac{mg}{f^2\mu}$.

a Plot the graph of the results (see the table in question **7**) of the independent and dependent variables (identified in question **6**) using the equation $\lambda^2 = \dfrac{mg}{f^2\mu}$.

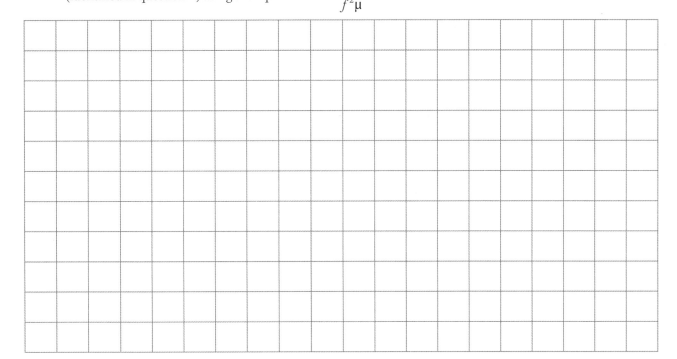

b What is the equation for the gradient using this formula?

_____

_____

c Find μ when the frequency of the vibration generator is set to 50 Hz.

_____

_____

d Why have we used λ^2 vs mass for the graph?

_____

_____

Quantitatively analysing the Doppler effect and beats

STUDENT BOOK
Pages 261–3

LEARNING GOALS

Understand the change in frequency of a wave due to the Doppler effect

Identify that the beat frequency is the difference between two frequencies

The superposition of waves gives rise to the phenomenon of beats through alternating constructive and destructive interference.

The rarefactions and compressions of sound waves contribute to a phenomenon called the Doppler effect, where the frequency of a sound appears to change depending on the velocity of the observer relative to the source of the sound. During this worksheet we are going to be looking at practising the quantitative analysis of the beats and Doppler equations.

The beats equation is shown as $f_{beat} = |f_2 - f_1|$ where f_1 and f_2 are two source frequencies.

The Doppler equation is shown as $f' = f\dfrac{(v_{wave} + v_{observer})}{(v_{wave} - v_{source})}$. This is reduced in the following situations.

Motion of the source of sound and the observer, and the relationship between the source frequency, f, and the observed frequency, f'.

Motion of observer	Motion of source	f' compared with f	Formula
Stationary	Towards observer	$f' > f$	$f' = \dfrac{v}{(v - v_s)}f$
Stationary	Away from observer	$f' < f$	$f' = \dfrac{v}{(v + v_s)}f$
Towards source	Stationary	$f' > f$	$f' = \dfrac{(v + v_o)}{v}f$
Away from source	Stationary	$f' < f$	$f' = \dfrac{(v - v_o)}{v}f$

1 Ayse is standing on the side of a road and a fire engine is approaching her. The fire engine is traveling at $60\,km\,h^{-1}$ and its siren has a frequency of $1450\,Hz$. Using the speed of sound in air as $340\,m\,s^{-1}$, what is the frequency Ayse hears when the fire engine is:

a approaching her

_____

_____

_____

_____

b moving away from her.

_____

_____

_____

_____

_____

2 Rocco and Serena are driving towards each other and both sound their horns. Both horns have a frequency of 182 Hz. Rocco is driving at $70\,km\,h^{-1}$. Serena is driving at $110\,km\,h^{-1}$. Assume the speed of sound in air is $340\,m\,s^{-1}$.

a What frequency does Serena's horn sound to Rocco?

_____

_____

_____

_____

b What frequency does Rocco's horn sound to Serena?

_____

_____

_____

_____

3 In a trainee orchestra, the violinists are supposed to be playing an $A_4 = 440\,Hz$ but one of the violinists is playing a note with a frequency of 431 Hz. What is the beat frequency?

_____

_____

_____

4 The bass guitarist and lead guitarist in a band are producing a beats frequency of 7 Hz. If the frequency of the note produced by the bass guitar is 390 Hz, what could be the possible frequencies of the note produced by the lead guitar?

_____

_____

_____

_____

INQUIRY QUESTION: WHAT PROPERTIES CAN BE DEMONSTRATED WHEN USING THE RAY MODEL OF LIGHT?

WS 10.1 **Modelling the properties of images formed in mirrors and lenses**

STUDENT BOOK
Pages 268–83

LEARNING GOALS

Use ray diagrams to predict positions of objects and images

Understand the differences between mirrors and lenses

1 Identify the lenses and mirrors shown below.

Converging lenses (thicker in the centre)

a   b   c   d   e k

_____ _____ _____ _____ _____

Diverging lenses (thinner in the centre)

f   g   h   i   j

_____ _____ _____ _____ _____

2 On the diagram below, label the blank spaces with the following labels:

Focal point, Principal axis, Focal length, Lens axis, Optical centre, L, F, f

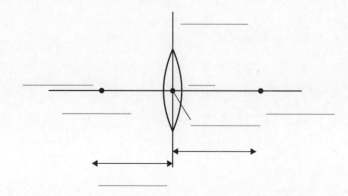

3 Draw a set of parallel rays travelling from a source and show what happens when they pass through the following lenses.

a A converging lens

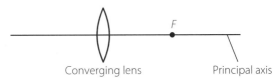

Converging lens   Principal axis

b A diverging lens

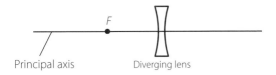

Principal axis   Diverging lens

Drawing ray diagrams with converging lenses

It can become confusing to show many different rays when we are drawing ray diagrams. As an alternative, we can use three significant rays that will allow us to find the properties of the images that are formed by the lens or mirror.

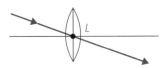
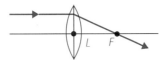
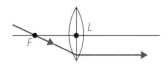

A ray through the optical centre *L* is undeviated.

A ray parallel to the principal axis is refracted (by the lens) to pass through *F*.

A ray arriving through *F* is refracted parallel to the principal axis.

When drawn to scale, an accurate ray diagram allows us to measure the object distances using a ruler.

> **HINT**
>
> The use of a sharp (fine pointed) pencil helps with the accuracy of ray diagrams.

4 a An 8.0 cm carrot is placed 25 cm in front of a converging lens with a focal length of 15.0 cm. Draw an accurate ray diagram of this set-up. When drawing a ray diagram choose a suitable scale so that it will fit in the available space.

Use the diagram to answer the following questions.

b Is the image real or virtual? Why?

_____

_____

c What is the size of the image?

_____

5 a A 5.0 cm high pencil is placed 4.0 cm away from a converging lens with a focal length of 7.0 cm. Draw an accurate ray diagram of this set-up.

Use the diagram to answer the following questions.

b Is the image real or virtual? Why?

_____

_____

c What is the size of the image?

_____

Drawing ray diagrams with diverging lenses

6 a Sebastian is looking into a diverging lens in a fairground funhouse. Sebastian is 178 cm tall and is standing at a distance of 50.0 cm away from the lens. The focal length for the lens is −25.0 cm. Draw an accurate ray diagram of this set-up.

Use the diagram to answer the following questions.

b What type of image is formed?

_____

_____

c What is the position of the image?

_____

_____

d What is the size of the image?

_____

Drawing ray diagrams with curved mirrors

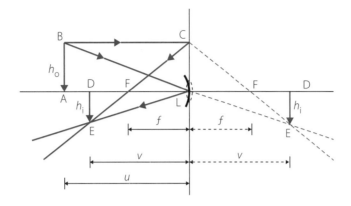

7 a A drink can that is 15 cm tall is seen in a concave mirror. It is 7.0 cm away from the mirror, which has a focal length of 5.0 cm. Draw an accurate ray diagram of this set-up.

Use the diagram to answer the following questions.

b What type of image is it?

_____

c What is the position of the image?

_____

d What is the size of the image?

_____

Drawing ray diagrams with diverging mirrors

The geometry for a diverging mirror will be the opposite to that of a diverging lens, as in the diagram below.

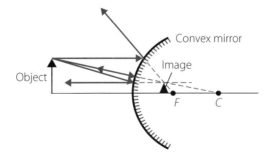

8 a Taneesha is walking towards a corner and there is a convex mirror of the corner to help see around it.
If Taneesha is 150.0 cm away from the mirror and is 169.0 cm tall, where would the image be if the focal length of the mirror is 70.0 cm? Draw an accurate ray diagram of this set-up.

Use the diagram to answer the following questions.

b Is the image real or virtual?

_____

c What is the size of the image?

_____

The thin lens equation

We have shown that you can use a ray model of light to model properties of an image, including where it is, whether it is real or inverted, and its magnification. Another way used in physics to find out this information is called the *thin lens equation*. This is not included in the syllabus, but it can be useful for trouble-shooting your answers when drawing a ray diagram. The thin lens equation is as follows:

$$\frac{1}{u} + \frac{1}{v} = \frac{1}{f}$$

where u is the distance from the object to the lens, v is the distance from the lens to the image, and f is the focal length of the lens.

Remember: this is extra information that will not be included on any data sheets, so you will need to remember it for an exam.

9 Use the thin lens equation to check whether the image in the ray diagram for question **4** above is real or virtual.

_____

_____

_____

_____

_____

_____

We can use the distances from the image and the object to work out the magnification of the image as well using the formula $M = -\frac{v}{u}$, where v is the distance to the image and u is the distance to the object. This is used when we do not have the height of the object or image. If we have the height of the object and image, we can use the relationship $M = \frac{h_i}{h_o}$, where h_i is the height of the image and h_o is the height of the object.

10 What is the size of the image in question **4**?

_____

_____

11 Is the image in question **5** real or virtual?

_____

_____

_____

12 What is the magnification of the image in question **5**?

_____

_____

13 What is the size of the image in question **5**?

_____

_____

We can predict pathways and angles of refraction using the following equations.

Refractive index: $n_x = \dfrac{c}{v_x}$

Snell's law: $n_1 \sin\theta_1 = n_2 \sin\theta_2$

Critical angle for a medium: $\sin\theta_c = \dfrac{n_2}{n_1}$; also written as $\sin\theta_c = \dfrac{1}{n_x}$ if $n_2 = 1$, which it does in air or a vacuum.

1 How can we tell which direction the light ray will refract when changing medium?

_____

_____

2 For the following problems, show the pathway of the light through the medium and work out the angle of refraction.

a $n_1 = 1.00$, $n_2 = 1.65$, $\theta_1 = 30°$, $\theta_2 = ?$

b As light passed through a medium it slowed down to $2.00 \times 10^8 \, \text{m s}^{-1}$. The angle of refraction is $18°$.

Using this information, work out:

i the refractive index of the medium

_____

_____

ii the angle of incidence

_____

_____

c The diagram below shows total internal reflection for a glass prism sitting in water. Work out the critical angle for total internal reflection to happen.

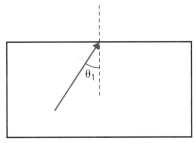

n_1 = glass 1.50
n_2 = water 1.33

_____

_____

3 Using your understanding of how light travels through glass, why would the NBN use fibre optic cables to transfer information as opposed to copper phone lines?

_____

_____

_____

_____

4 How does the composition of a fibre optic cable allow it to transfer light along it?

_____

_____

_____

_____

Identify that light will follow the inverse square law

Use the inverse square law to predict light intensity at different distances

As light leaves a point source, its intensity decreases in line with the inverse square law. The inverse square law also applies to sound (see chapter 9).

1 What is your understanding of the inverse square law?

_____

_____

_____

_____

_____

_____

INVESTIGATION

The inverse square law

A group of students want to show how the inverse square law works with light.

AIM

To measure the variation in light intensity with the distance from the source

2 Write a suitable hypothesis for this experiment.

_____

_____

MATERIALS

- Large cardboard box • Mini Maglite® torch (or similar) with the cover taken off (this will just have the bulb)
- Graph paper (squares 0.5 cm wide) • Separate sheet of cardboard • Ruler

METHOD

1 Take the top off your cardboard box.
2 Stick the graph paper to the bottom of the box and turn the box on its side.
3 Make a 10-cm long handle for the torch.
4 Cut a 1 cm × 1 cm hole in the plain sheet of cardboard.
5 Stick the handle to this piece of card and mount the torch (see diagram on the next page).
6 Measure the number of squares on the graph paper illuminated by the light and record this in the results table.
7 Move the cardboard back away from the graph paper at regular intervals, making sure that you measure from the bulb and not the card.
8 Repeat the above steps 10 times.

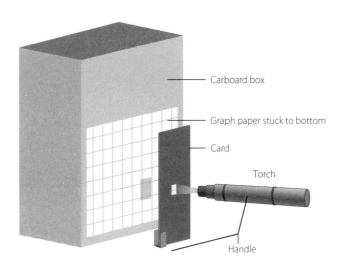

- Carboard box
- Graph paper stuck to bottom
- Card
- Torch
- Handle

RESULTS

The squares on the graph paper are 0.5 cm across, giving them an area of 0.25 cm^2.

To work out relative brightness we use the equations $B = \dfrac{L}{A}$ where B = brightness at any distance L, and $B_0 = \dfrac{L}{A_0}$ for

the standard distance of 10 cm. This rearranges to give $\dfrac{B}{B_0} = \dfrac{A_0}{A}$. A_0 will cover 1 cm^2 at 10 cm so $A_0 = 1$, making the

equation for relative brightness $\dfrac{B}{B_0} = \dfrac{1}{A}$.

3 a The table below shows the results. Fill in the empty rows.

Distance from bulb (cm)	Number of squares illuminated	Area illuminated (cm^2)	Relative brightness (cm^{-2})
10	4		
14	8.4		
15	9.3		
18	13.3		
20	16.4		
24	23.5		
25	26		
28	34.8		
30	36.6		

b Choose a suitable scale and plot the distance (on *x*-axis) against relative brightness (on *y*-axis) on the grid below.

DISCUSSION

4 Do the results of the investigation support the hypothesis?

_____

5 Why did the method use a holder at a set distance from the hole in the cardboard?

_____

_____

_____

11 Thermodynamics

WS 11.1 Measuring temperature

LEARNING GOALS

Define temperature

Explain the relationship between kinetic energy and temperature

Identify the SI unit for temperature

1 Define 'temperature'.

_____

2 Describe how the movement of particles within an object changes as the temperature of the object increases.

_____

_____

_____

Kelvin (K) is the SI unit for temperature. The scale has a base point of 0 K, also known as absolute zero. An object at 0 K has all its particles in the ground (lowest energy) vibrational state and its temperature cannot be reduced any further.

3 Outline the Celsius scale of temperature and explain why in specific situations Celsius and Kelvin are interchangeable.

_____

_____

_____

_____

_____

_____

4 Why is it not practical to measure the temperature in the upper atmosphere (where the air density is extremely low)?

_____

_____

_____

_____

_____

5 a Particles in liquid or gaseous states can exhibit kinetic energy in other modes in addition to vibrational. Describe two of these modes.

_____

_____

_____

b Why are these modes not evident in solids?

_____

_____

_____

6 Does $1\,m^3$ of steel at $300\,K$ contain the same amount of heat energy as $1\,m^3$ of air at $300\,K$? Explain.

_____

_____

_____

_____

_____

_____

LEARNING GOALS

Define equilibrium

Explain how equilibrium relates to the movement of heat energy

Calculate temperature and mass in thermal equilibrium problems

1 Define 'equilibrium'.

2 How can this definition be applied to the concept of heat energy?

When two objects are placed in thermal contact with each other they will exchange thermal energy in order to achieve thermal equilibrium. The law of conservation of energy dictates that the amount of thermal energy gained by one object is the same as the energy lost by the other object in an isolated system. If we have two objects consisting of identical substances with known masses, we can calculate the changes in temperature by using the conservation of energy. The temperature at equilibrium multiplied by the total mass is equal to the sum of the temperatures multiplied by the respective masses of the constituents according to the following equation.

$$T_t m_t = m_1 T_1 + m_2 T_2$$

This equation is not on the formula sheet but is derived from a weighted average calculation.

3 1 kg of water at 73°C was mixed with 1 kg of water at 25°C. What is the final temperature of the water?

4 When 15.4 kg of coolant at 84°C was cooled by mixing it with 10.0 kg of chilled coolant, the final temperature of the coolant was found to be 53.2°C. What was the initial temperature of the chilled coolant?

5 What mass of octane at room temperature (25°C) would be required to cool 7.5 kg of octane from 80°C to 40°C?

6 A mass of glycerol, A, was found to be at 22.45°C. When an equal mass of glycerol, B, was added, the temperature rose by 15.35°C. What was the initial temperature of glycerol B?

Analysing the relationship between specific heat capacity and temperature change

LEARNING GOALS

Define specific heat

Explain the origins of specific heat

Calculate the thermal energy obtained or released by a substance

The relationships seen in worksheet 11.2 are only relevant when considering identical substances. Different substances (even with the same mass), and even different states of the same substance, require different amounts of energy in order for their temperature to increase by 1°C or 1K. (Remember these are interchangeable when they refer to a change in temperature.)

To compare different substances we must consider their *specific heat capacity*, which is defined as the energy required to raise the temperature of 1 kg of a substance by 1 K and therefore has units of $J\,kg^{-1}\,K^{-1}$. The value for liquid water is $4.18 \times 10^3\,J\,kg^{-1}\,K^{-1}$.

Water has a very high heat capacity when compared to most other substances. The table below lists the specific heat capacities for common substances.

Substance	Specific heat capacity ($\times 10^3\,J\,kg^{-1}\,K^{-1}$)
Water	4.18
Ice	2.03
Steam	2.08
Air	1.01
Salt	0.864
Aluminium	0.902
Iron	0.450
Copper	0.385
Gold	0.129

State your answers to an appropriate number of significant figures.

1 What causes different substances (and different states of the same substance) to have different specific heat capacities?

_____

_____

_____

2 Explain the reasoning behind using water as a coolant.

_____

_____

_____

3 Some solar electricity systems (called solar thermal powerplants) use mirrors to concentrate light on a large water tank. This heats the water to turn it into steam to drive a turbine. What limitation does water have in this application?

_____

_____

_____

4 Lydia left a beaker of water and a copper cube (of the same mass) next to each other in direct sunlight at the start of the school day. When she measured the temperature of each using a laser thermometer 4 hours later, she noted that the surface temperature of the copper was much higher than that of the water. Explain this observation.

_____

_____

From the definition on page 156, we derive the equation $Q = mc\Delta T$, where ΔQ is the heat transferred to the material, m is the mass of the material, ΔT is the temperature change and c is the specific heat capacity of the material.

5 How much thermal energy would be required to raise the temperature of 200 g of water by 7.5 K?

_____

_____

6 8.46×10^4 J of energy was delivered to 2000 g of copper at 20°C. What is the final temperature of the copper?

_____

_____

_____

7 A kettle delivered 4.72×10^4 J of energy to an unknown volume of water and raised its temperature by 78°C. How much water was in the kettle?

_____

_____

8 A 735 g block of metal was measured to have a temperature of 198°C. It was then dropped into a beaker containing 240 g of water at 23.5°C. After the system had reached equilibrium, the water temperature was 37.8°C. What is the specific heat capacity of the metal?

_____

_____

_____

_____

HINT
First, we need to consider the thermal energy gained by the water.

9 A student suggested using an oil ($c = 2.09 \times 10^3$ J kg^{-1} K^{-1}) instead of water as a solution to the problem encountered in question **3**. Why is this not a viable option from a thermodynamics point of view?

_____

_____

_____

HINT
The boiling point of the oil is 278°C.

Explain change of phase

Define latent heat

Write a research question

Complete a risk assessment

Determine thermal transfer rate

Graph variables

Determine a relationship from a graph

When we transfer thermal energy from a source to water, we expect the temperature to rise as modelled by $Q = mc\Delta T$. However, when the water begins to boil or vaporise (change state), the temperature of the water does not rise even though more heat energy is being transferred to the water. This extra energy is used to change the state of water from liquid to gas and is referred to as the *latent heat of vaporisation*. Latent heat is constant for a given phase change of a given substance. Because there is no temperature change, latent heat of change of state has units of $J\,kg^{-1}$. The latent heat of vaporisation of water is $2.26 \times 10^6\,J\,kg^{-1}$.

1 Explain what it means for a substance to undergo a change of state.

_____

_____

_____

_____

2 Define 'latent heat'.

_____

_____

3 Compare latent heat of vaporisation with latent heat of fusion.

_____

_____

_____

_____

_____

_____

INVESTIGATION

Latent heat of vaporisation of water

AIM

To determine the latent heat of vaporisation of water

RESEARCH QUESTION

4 Write a research question for this experiment.

_____

_____

5 Complete the following risk assessment table for this investigation.

What are the risks in doing this investigation?	How can you manage these risks to stay safe?
Burns from hot plate, hot water or hot glassware	Use protective equipment or tongs when handling hot equipment
	Turn off hotplate immediately after vaporisation is complete

MATERIALS

- Temperature-controlled hotplate
- Thermometer
- 50 mL beaker
- 100 mL distilled or deionised water
- Stopwatch
- Electronic balance

METHOD

1 Place 10 mL of water in a 50 mL beaker.

2 Measure and record the initial temperature of the water using the thermometer.

3 Turn on the hot plate to its highest setting and allow it to warm up for 3 minutes.

4 Place the 250 mL beaker of water on the hotplate.

5 Use the stopwatch to measure the time taken for the temperature to increase by 10 K and record the time in Table 1 below.

6 Continue heating the water and start timing once boiling has commenced.

7 In table 2, record the time it takes for all the water to vaporise. Be sure to turn off the hotplate immediately after vaporisation has completed.

8 Repeat steps 6 and 7 with 15, 20 and 25 mL of water.

RESULTS

6 Calculate the total thermal energy transfer to the water to raise its temperature by 10 K using $Q = mc\Delta T$. Remember, the specific heat capacity of water is $4.18 \times 10^3 \, \text{J K}^{-1} \text{kg}^{-1}$.

_____

_____

7 Calculate the rate at which heat is transferred to the water.

_____

_____

Table 1 Rate of heat transfer to the water

Time (s)	Heat transferred (J)	Rate of transfer (J s^{-1})

To find the heat transferred to each mass of water while it is vaporising, we take the value for the rate from the above table and multiply it by the time taken to vaporise each mass of water.

Table 2 Heat required to vaporise water

Mass of water (kg)	Time (s)	Heat transferred (J)
0.01		
0.015		
0.02		
0.025		

8 Create a graph of Q as a function of mass of water below.

We know that the latent heat of change of state has units of $J\,kg^{-1}$, which are the same as the units of the gradient of this graph.

9 Determine a value for the latent heat of vaporisation of water from the gradient of your graph.

_____

_____

_____

DISCUSSION

10 How does your value compare to the accepted value of $2.26 \times 10^6\,J\,kg^{-1}$?

_____

_____

_____

11 Discuss the accuracy, reliability and validity of this investigation in determining the latent heat of vaporisation.

_____

_____

_____

_____

_____

_____

_____

_____

_____

_____

_____

12 If you were going to carry out this investigation again, how would you improve the above?

_____

_____

_____

_____

CONCLUSION

13 Have the aim and research question been addressed? Write a suitable conclusion for your investigation.

_____

_____

_____

Define thermal conductivity

Compare thermal conductivity and specific heat

Calculate thermal conductivity

Relate thermal conductivity to the uses of substances

State your answers to an appropriate number of significant figures.

1 Define thermal conductivity.

_____

2 Using $\dfrac{Q}{t} = \dfrac{kA\Delta T}{d}$, determine the units of thermal conductivity k.

_____

3 Compare the meaning of the thermal conductivity constant k and the meaning of specific heat capacity c.

_____
_____
_____

_____
_____

4 Why does thermal conductivity depend on cross-sectional area, not just length of material?

_____
_____
_____

5 How would the thermal conductivity constant of a material determine some of its potential uses?

_____
_____
_____

6 Why would it be difficult to determine the thermal conductivity of a fluid such a water or air?

_____
_____
_____

7 A CPU heat sink inside a computer is required to conduct 120 J of thermal energy each second away from a CPU that runs 10°C hotter than the surrounding air. Given the sink's conductive area is $16\,cm^2$ and it is 4 mm thick, what would be the minimum thermal conductivity needed for the heat sink to meet requirements?

8 a To deal with the rising temperatures in Australia, Jamal decided to reinsulate his bedroom. The bedroom has an external wall of surface area $17\,m^2$ and thickness of 65 mm, and it is made of a material with thermal conductivity of $0.005\,65\,W\,m^{-1}\,K^{-1}$. Calculate how long it would take to raise the internal temperature by 5°C when the external temperature is 25°C higher than the internal temperature. Assume the room is empty and filled with 31.25 kg of air, which has a specific heat capacity of $718\,J\,kg^{-1}\,K^{-1}$.

b How would the time required for the temperature to increase by this amount change if the thickness of the insulative material was tripled?

Module three: Checking understanding

Circle the correct answer for questions 1–6.

1 Which wave has the longest wavelength?

  A An infrared wave with a frequency of 7.0×10^{10} Hz

  B A sound wave of frequency 17 kHz

  C A water wave with a frequency of 0.25 Hz travelling at $7.0 \, \text{m s}^{-1}$

  D A 150 MHz radio wave

2 Looking at the statements below, which one is always true for refraction when a light wave travels between mediums?

  A Total internal reflection will occur if the angle of incidence is suitable.

  B The light will bend towards the normal.

  C The wavelength of the light will change.

  D The speed of light will remain constant.

3 If a ray of light travels from the air into a Perspex rectangle at an angle of 35°, what is the refracted angle when the refractive index given for air is 1 and Perspex is 1.52?

  A 0.33°

  B 3.04°

  C 19°

  D 22°

4 When using the ray model to find an image, how do you know if the image is real?

  A The ray lines cross at a single point.

  B The ray lines run parallel to each other.

  C The ray lines are only reflected.

  D The ray lines are only refracted.

5 The resultant wave that has undergone superposition:

  A is smaller than the contributing waves

  B is the sum of the contributing waves

  C is the same size, because superposition has no net effect

  D has an increase in its frequency

6 Petra placed 450 g of water at 22°C in the freezer with a digital thermometer attached. She observed the temperature as the water went from 22°C to −6°C.

  a Draw the cooling graph for the water over time.

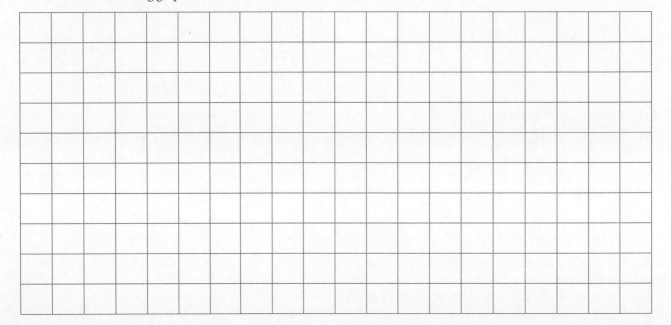

b Explain the considerations you used to construct the graph.

_____

_____

_____

_____

7 A 20 kg block of aluminium is at 85°C. It is placed in a vat that contains 50 kg of water at 15°C.

Data:

Aluminium: $c_{Al} = 900 \, \text{J kg}^{-1} \text{K}^{-1}$

Water: $c_w = 4200 \, \text{J kg}^{-1} \text{K}^{-1}$

Calculate the temperature at which the aluminium and water will come to thermal equilibrium.

_____

_____

_____

MODULE FOUR »
ELECTRICITY AND MAGNETISM

Reviewing prior knowledge

1 Identify the two types of electric charge and the particles responsible for them.

2 Explain the interactions between two like and two unlike charged particles.

3 Define 'voltage', 'current' and 'resistance'.

4 Identify the relationship known as Ohm's law.

9780170449595

5 Define electrical 'conductor' and 'insulator' and provide two examples of each.

_____

_____

_____

_____

_____

_____

6 Explain the difference between series and parallel circuits.

_____

_____

_____

_____

7 Identify three magnetic materials.

_____

_____

_____

8 a Outline the interactions between two north poles.

_____

_____

b Outline the interactions between a north and a south pole.

_____

_____

9 Explain how a navigation compass works.

_____

_____

_____

_____

Electrostatics

WS 12.1 Charge and the electrostatic force

STUDENT BOOK
Pages 337–43

LEARNING GOALS

Explain how a net charge is produced

Evaluate experimental accuracy

Write a scientific procedure

Charge is an intrinsic property of matter. Charge allows particles to experience the electrostatic force. Most objects have equal numbers of positive and negative charges and thus have no net charge. If there is an imbalance between positive and negative charges within an object, the object is considered to be 'charged'.

1 Explain the process by which objects become electrically charged.

_____

_____

_____

_____

_____

_____

_____

2 Outline the electrostatic force and describe how charges interact.

_____

_____

_____

_____

_____

_____

_____

3 Olivia and Sabina conducted an investigation to observe and qualitatively determine the relationship between the variables affecting the electrostatic force between charged objects. Their findings are shown below.

Comment on the accuracy of each of the conclusions provided by the students.

Variable	Change	Observation	Conclusion
Charge type	Positive and positive	Objects repelled each other	Like charges repel and opposite charges attract
	Negative and negative	Objects repelled each other	
	Positive and negative	Objects were attracted to each other	
Separation distance	Increased	Only small mass objects were attracted	Electrostatic force is inversely proportional to the separation distance between charges
	Decreased	Bigger mass objects were attracted	
Charge magnitude	Increased	Bigger mass objects were attracted	The magnitude of the electrostatic force is proportional to the magnitude of the charge
	Decreased	Only small mass objects were attracted	

4 Suggest a procedure by which one of the relationships described in question **3** could be investigated quantitatively.

LEARNING GOALS

Outline equipotential lines

Describe the purpose of field line diagrams

Graphically represent electric fields

The electrostatic force is a field force much like the gravitational force. This means we can also model electric fields using field lines. As with other field forces, the field line arrows point in the direction of the field and the separation distance between lines indicates the relative strength of the field.

1 How is the direction of an electric field defined?

_____

2 What does the relative spacing of field lines represent?

_____

_____

3 Why can field lines not cross?

_____

_____

4 What is the relationship between electric field lines and lines of equipotential?

_____

_____

_____

5 Draw the electric field lines for the following situations.

a   b

+   −

c   d

+   +   −   −

e

+   −

f

2+   −

g

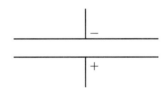

2+   +

h

+

−

i

−

+

6 Explain the difference between the diagrams drawn in question **5e** and **f**.

_____

_____

_____

7 The outermost field lines between parallel charged plates bend outwards, whereas all the other field lines are perpendicular to the plates. Why is this so?

_____

_____

_____

_____

Applying the electric field model

Derive an expression for force on a charge due to an electric field

Apply mathematical models to problems involving electric fields

The force between spherical and plate charges and their respective fields can be modelled quantitatively through the use of a number of equations.

The electric field for a point charge can be described as radial, with a field strength proportional to field line density. This tells us that $\vec{E} \propto \dfrac{q}{r^2}$. The constant required to turn this relationship into an equation is the constant of proportionality $\dfrac{1}{4\pi\varepsilon_0}$, where ε_0 is the permittivity of free space.

$$\varepsilon_0 = 8.858 \times 10^{-12}\,\mathrm{C^2\,N^{-1}\,m^{-2}}$$

When we consider the electric field at $1\,\mathrm{m}$ from a $1\,\mathrm{C}$ point-like charge, we get

$$\vec{E} = \frac{1}{4\pi(8.9 \times 10^{-12}\,\mathrm{C^2\,N^{-1}\,m^{-2}})} \times \frac{1\,\mathrm{C}}{1\,\mathrm{m}^2} = 9 \times 10^9\,\mathrm{N\,C^{-1}}$$

The units obtained here are in the form of force per charge. Hence we can say that $\vec{E} = \dfrac{\vec{F}}{q}$, or $\vec{F} = \vec{E}q$.

Remember that the fundamental charge has a magnitude of $1.602 \times 10^{-19}\,\mathrm{C}$.

State your answers to an appropriate number of significant figures.

1   An electron is placed in a uniform electric field of $4.0 \times 10^3\,\mathrm{N\,C^{-1}}$ directed up the page. What is the force acting on the electron due to the field?

_____

_____

2   What field strength and direction would cause a proton to experience a force of $6.42 \times 10^{-11}\,\mathrm{N}$ from left to right of the page?

_____

3   An electric field is used to separate alpha (charge $3.204 \times 10^{-19}\,\mathrm{C}$) and beta (charge -1.602×10^{-19}) particles during radioactive decay. Determine the magnitude and direction of the force each particle will experience in a field of $1.00 \times 10^5\,\mathrm{N\,C^{-1}}$ to the right.

_____

_____

4   Calculate the acceleration experienced by an electron when it is placed in a field of strength $465\,\mathrm{N\,C^{-1}}$ created by two parallel plates.

_____

_____

_____

By combining $\vec{E} = \dfrac{1}{4\pi\varepsilon_0}\dfrac{q}{r^2}$ and $\vec{F} = \vec{E}q$, we can develop an expression for the force between two point-like charges.

$$\vec{F} = \dfrac{1}{4\pi\varepsilon_0}\dfrac{q_1 q_2}{r^2}$$

5 Two charges experience a force F when placed a distance d apart. What is the force when the charges are moved to a separation of $\dfrac{d}{4}$?

_____

_____

_____

_____

6 Two charges, each of charge q, are placed at a distance d from each other. If the charges were doubled, what relative separation distance would provide the same force?

_____

_____

_____

7 What force is experienced by an electron when placed 24 mm from a point-like charge of 3.942 nC?

_____

_____

_____

_____

8 A proton is placed 12 cm above a point-like charge so that the down force due to gravity is precisely cancelled out by the electrostatic repulsion. What is the magnitude and sign of the point charge?

_____

_____

_____

_____

_____

9 Two point-like charges each have charge 1.5×10^{-7} C. What is the closest distance the two spheres can be without the force being greater than 7.4 μN?

_____

_____

_____

To find the electric field between two parallel plates, we divide the potential difference between the two plates by the separation distance: $\vec{E} = \dfrac{V}{d}$. Note that this shows the electric field strength can have units of volts per metre (V m^{-1}).

10 Two charged parallel plates produce an electric field strength of $400\,\text{N C}^{-1}$ when a potential difference of $12\,\text{V}$ is applied across the gap. What is the separation distance between the two plates?

_____

_____

_____

11 A pair of parallel plates have a separation distance of $5\,\text{mm}$. What voltage would be required to produce a field strength of $1200\,\text{N C}^{-1}$ in the region between the plates?

_____

_____

12 A capacitor with plate separation of $20\,\mu\text{m}$ ($2 \times 10^{-6}\,\text{m}$) is placed in a $5\,\text{V}$ circuit. What is the electric field strength between the two plates?

_____

_____

13 Compare the equations for force produced by a point charge and force produced by a set of parallel charged plates. How does doubling the separation distance affect each of these situations? Explain why.

_____

_____

_____

_____

_____

Analysing the effects of a moving charge in an electric field

STUDENT BOOK
Pages 355–64

LEARNING GOALS

Use the concept of work to relate electric and dynamic properties of particles
Apply mathematical models to problems involving work

As we have learnt, objects with mass placed in a gravitational field have potential energy known as gravitational potential energy. Likewise, electrical charge has potential energy when placed in an electric field. This means that work is done when a charge moves or is moved to a position of different potential energy within an electric field.

1 Develop an expression for the work done by an electric field on a charge within the field.

_____

_____

_____

When considering the conservation of energy, any work done on a charge in an electric field will equal a change in its potential energy. We can say that $W = Eqd = -\Delta U$, noting that positive work done by the field will reduce the potential energy and negative work down will increase the potential energy.

2 Write an expression relating the potential difference (in volts) between two points within an electric field and the change in potential energy of a charge that moves between these points.

_____

_____

_____

3 Recall what an 'equipotential line' is.

_____

_____

4 Using the equation from question **2**, describe the work done when a charge moves along a equipotential line.

_____

_____

5 Calculate the change in potential energy when an electron moves 5.0 mm within a uniform electric field towards the positive plate with an electric field strength of $2.54 \times 10^{-6} \, \mathrm{N\,C^{-1}}$.

_____

_____

_____

6 A proton has $8.19 \times 10^{-16} \, \mathrm{J}$ of energy transferred to kinetic energy as it moves from point A to point B within a uniform electric field. What potential difference does this relate to?

_____

_____

_____

7 Show that the change in potential energy of a point-like charge in a uniform electric field can be written with units of electron volts (eV).

_____

_____

_____

_____

_____

When positive work is done on a charge within an electric field, the charge will accelerate as potential energy is converted to kinetic energy (once again, think conservation of energy). We can now equate a change in kinetic energy with electrical potential energy $Eqd = \Delta E_k = \dfrac{1}{2} mv^2$.

8 A proton travelling at $2.5 \times 10^6 \, \text{m s}^{-1}$ enters an electric field of strength $1.0 \times 10^4 \, \text{V m}^{-1}$. The field lines of the electric field are in the opposite direction to the proton's velocity. How far will the proton travel in the field before it stops?

_____

_____

_____

_____

_____

_____

9 Calculate the voltage required to levitate a proton between two parallel charged plates spaced 1.00 mm apart.

> **HINT**
>
> Equate the electrical force to the gravitational force.

_____

_____

_____

_____

_____

_____

13 Electric circuits

WS 13.1 Applying models to represent current in metals

STUDENT BOOK
Pages 369–73

LEARNING GOALS

Identify the electrical properties of common materials

Define current

Apply mathematical models to problems involving charge and current

Write a conclusion

State your answers to an appropriate number of significant figures.

1 Materials can be classified based on how easy it is for a charged particle to pass through them. Explain the properties of conductors and insulators.

2 Define 'current'.

3 What is the relationship between charge passing through a material and current?

4 A current of 1 A was measured in a simple circuit. How many electrons must have passed a point each second for that reading?

5 A current of 0.45 A flowed through an electrical component over a 72 second period. Calculate the total charge that passed through the component.

6 A battery contained a total of 7.62×10^6 C of charge. If it was connected to a device that drew 1.60 A, how long would the battery be able to supply the device with charge (assuming the battery could fully discharge)?

_____

7 A mobile phone battery has a total charge capacity of 2.3×10^3 C. If it is charged from empty to full in 42 minutes, what is the current supplied to the phone?

_____

_____

8 Rima and Aishani investigated the rate at which charge passed through a variety of conductors for a constant voltage. To do this they set up a circuit containing a power supply, a resistor, an ammeter and the sample of each metal (with consistent cross-sectional area and length). Their results are shown below.

Metal sample	Current (A)
Iron	0.87
Aluminium	1.0
Copper	1.2
Magnesium	0.93

a Why would it be suitable to measure the current passing through the sample to determine the rate at which charge flows through the sample?

_____

_____

b What can you qualitatively conclude about how quickly charge can move through each of the metals that were tested?

_____

_____

9 The battery for a power tool is quoted as having a capacity of 5.2 amp hour. Explain what this means and calculate the total charge held by the battery.

_____

_____

_____

_____

_____

_____

9780170449595

LEARNING GOALS

Apply Ohm's law to mathematical problems

Graph experimental results

Determine experimental relationships

Ohm's law is very useful for determining unknowns for a specific electrical component and can also be helpful in determining the amount of energy that has been transformed by a component.

State your answers to an appropriate number of significant figures.

1 A fan motor has an internal resistance of $425\,\Omega$. What is the maximum current it can draw from a 240V power source at start up?

_____

2 What voltage drop will draw a current of 1.00A through a $1.465\,k\Omega$ resistor?

_____

_____

3 A piece of old circuitry was tested and found to allow a current of 7.2A to flow through it when a potential difference of 110V was applied across it. What is the resistance of the piece of circuitry?

_____

4 A computer component with a maximum current rating of 0.15A is connected in series with a 12V power source and a resistor. What resistance would be required in the resistor to prevent the component from burning out, assuming the component has zero resistance?

_____

5 a An old office building is to be rewired with new cabling. The packaging claims the cables have 20% less resistance than the older cables. If the older cabling allowed a current of 9.6A to flow when a 240V potential difference was applied, what resistance would the new cabling have?

_____

_____

b What is the difference in resistance between the two types of cabling?

_____

6 How much work is done on an electron when it is moved through a potential difference of 240V? Give your answer in both joules and electron volts.

_____

_____

_____

7 24.6eV of work are done in moving an electron between two charged parallel plates. What is the potential difference between the two plates?

_____

8 How much work is done in moving a charge of 2.4C through a potential difference of 18V?

_____

9 What charge would pass through a potential difference of 24 V if a total of 725 J of work is done on the electrons?

_____

_____

10 74.52 J of work is done on a charge of 1.84×10^{-2} C as it moves through a potential difference. What is the potential difference?

_____

_____

We have learnt about the relationship between voltage, current and resistance as Ohm's law. While Ohm's law applies when all other variables are controlled, resistance itself is also dependent on other factors, most notably temperature and frequency. As a result, electronic components can be classified as ohmic (where their resistance is not significantly affected by other factors like temperature) or non-ohmic (where the resistance is significantly determined by other factors and as a result $\frac{V}{I}$ does not provide a constant value). Simply put, the resistance is constant in ohmic components and variable in non-ohmic components.

11 Aaron and Reza conducted an investigation to classify three different components as ohmic or non-ohmic. To conduct this investigation, they used a circuitry resistor, an incandescent light globe and a length of nichrome wire. Each component was connected to a variable power supply and an ammeter, and the current was measured for a range of applied voltages. Their results tables are shown below.

Voltage (V)	Current (A)		
	Circuitry resistor	Incandescent light globe	Nichrome wire
2.0	0.20	1.3	1.5
4.0	0.40	1.6	2.2
6.0	0.60	1.9	2.7
8.0	0.80	2.1	3.2
10	1.0	2.3	3.6
12	1.2	2.4	4.0

    a Create a graph below of voltage as a function of current for each of the components used in the investigation.

b Using a linear regression line for each graph, classify each of the components as ohmic or non-ohmic based on the line or curve of best fit.

_____

_____

c Suggest a cause of the behaviour for the components that were classified as non-ohmic.

_____

_____

_____

12 Reza wanted to test a further component, a diode from an old radio. (A diode is a non-ohmic component made from semiconductors; your home lighting may use light-emitting diodes or LEDs.) His results are graphed below.

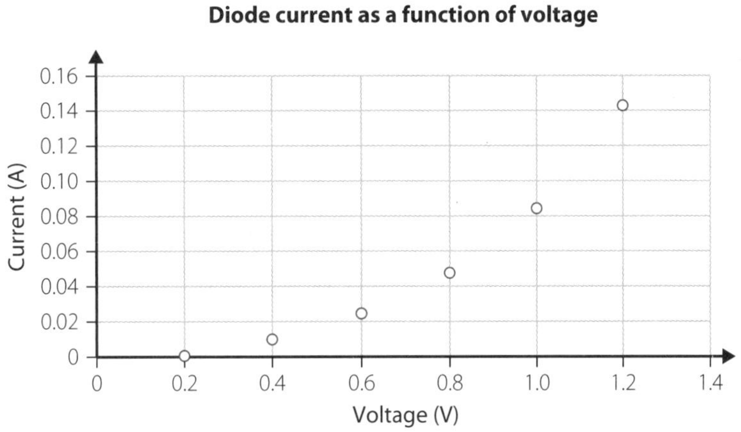

Diode current as a function of voltage

a Suggest what kind of relationship there is between current and voltage in the diode used for the test.

_____

b Extrapolating from the above graph, what might happen to the measurable current as the voltage drops to a non-zero value below 0.2 V?

_____

c What can you conclude about the usefulness of Ohm's law in modelling the resistance of components in a circuit?

_____

_____

_____

_____

_____

Converting electrical energy into heat and light

SB
STUDENT BOOK
Pages 380–5

State your answers to an appropriate number of significant figures.

1 Using your knowledge of power and work, derive an equation for power supplied to an electric circuit.

_____

2 An electric motor has a power rating of 145 W. Assuming 100% efficiency, how much charge must pass through the motor per second given a potential difference of 4.2 V?

_____

3 An incandescent light globe rated at 65 W is known to convert 12% of the electrical energy into light, with the remaining 88% being converted into heat. How much heat energy is produced if the light is operating for 2.4 h?

_____

_____

4 An electric kettle raised the temperature of 875 mL of water from 25°C to 98°C in 2.33 minutes. Calculate the power of the kettle.

_____

_____

_____

5 The kettle in question **4** is connected to a 240 V circuit with a load limit of 10 A. Will the kettle overload the circuit?

_____

6 Sarah and Connor conducted an investigation to determine the rate at which a coil of nichrome wire converts electrical energy into heat energy. They did this by setting up a simple kettle circuit.

a Write a suitable method that the students could have used to conduct this investigation.

_____

_____

_____

_____

_____

_____

_____

_____

_____

_____

The students' results are shown below.

Controlled variables	
Voltage	24 V
Current	5.2 A
Mass of water	0.200 kg

Time (s)	Temperature (°C)
0	22
60	28
120	35
180	43
240	52
300	61
360	74
420	82
480	88

Water temperature as a function of time

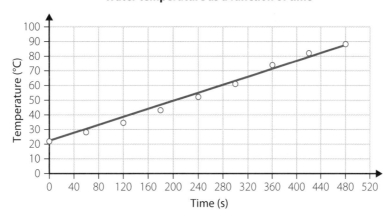

b Calculate the total electrical energy converted by the nichrome wire.

_____

c Calculate the total heat energy absorbed by the water (recall what you learnt in chapter 11).

_____

_____

d Explain any discrepancies between the values calculated above.

_____

_____

_____

_____

_____

_____

e How could Sarah and Connor make their experiment more reliable?

_____

_____

f Suggest three other devices that convert electrical energy into other forms of energy and identify the main form(s) of energy that are produced.

_____

_____

_____

7 Define 'energy efficiency'.

_____

_____

_____

8 Electrical and internal combustion vehicles operate by converting one type of energy into kinetic energy. Most internal combustion vehicles have efficiencies around 35%, meaning only 35% of the chemical energy in their fuel is converted into kinetic energy. Electric vehicles have much higher efficiencies (up to 90%). Explain these efficiencies in terms of the energy conservation laws.

_____

_____

_____

_____

_____

_____

_____

9 An air conditioner is capable of removing 5.54×10^6 J of energy from a room by consuming only 2.75×10^4 J of energy. Explain how this does not violate conservation laws.

_____

_____

_____

_____

_____

_____

WS 13.4 Conservation of charge and energy in complex circuits

STUDENT BOOK
Pages 386–97

LEARNING GOALS

- Apply conservation of charge to circuits
- Apply mathematical models to problems involving series and parallel circuits

Throughout this topic we have seen that Ohm's law can be used to model the relationship between voltage, current and resistance with respect to a single electrical component or resistor. Things become more complicated when resistors are added in parallel or when circuits have combined resistors in both parallel and series configurations.

To determine the total resistance within a complex circuit or the unknown resistance of a component, we need to consider how energy and charge are conserved within the circuit.

First, we must consider the conservation of energy. The potential energy of the charge moving within the circuit changes when it passes through a component such as a battery or a resistor. The potential energy increases as the charge passes through a power source such as a battery, and decreases when the charge passes through a resistor. The net change in energy within the circuit must be equal to zero.

We can summarise this as:

$$\Delta V_{tot} = V_{power\ source} - (V_{R_1} + V_{R_2} + ... + V_{R_n}) = 0$$

Using the law of conservation of energy, we can determine the potential difference across a component if the potential difference across each of the other components is known.

We also need to consider the conservation of charge, because no charge is removed or added to a circuit: energy is simply transferred between electrical and other forms. If we consider charge moving from one terminal of a battery to the other as positive, and charge moving through a resistor as negative, we can represent this as $\sum I = 0$. For circuits with parallel resistors, the current through each can be written as:

$$I_{power\ source} - (I_{R_1} + I_{R_2} + ... + I_{R_n}) = 0$$

State your answers to an appropriate number of significant figures.

1 Using the above conservation laws, write an expression for the total resistance of a series circuit.

_____

_____

_____

_____

2 Write an expression for the total resistance of a parallel section of a circuit.

_____

_____

_____

_____

3 Given the resistance of each individual resistor, what is the total equivalent resistance of the circuit below?

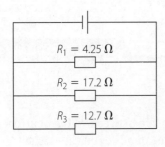

_____

_____

_____

_____

_____

4 Calculate the total equivalent resistance for the circuit drawn below.

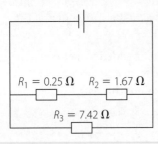

_____

_____

_____

_____

_____

5 Calculate the vale of R_1 in the circuit below, given the total equivlent resistance is 16.7 Ω.

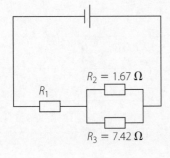

_____

_____

_____

_____

_____

_____

6 If a 12 V potential difference was applied to the following circuit, what current would it draw?

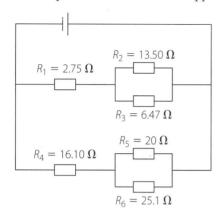

_____

_____

_____

_____

_____

_____

_____

_____

7 Calculate the resistance in R_1, given the total equivalent resistance of the circuit is $5.2\,\Omega$.

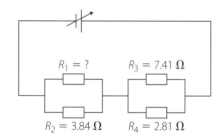

_____

_____

_____

_____

_____

_____

8 Calculate the value of R_3 in the following circuit, given the total equivalent resistance of the circuit is 47 Ω.

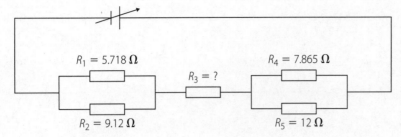

$R_1 = 5.718\ \Omega$

$R_3 = ?$

$R_4 = 7.865\ \Omega$

$R_2 = 9.12\ \Omega$

$R_5 = 12\ \Omega$

Magnetism

WS 14.1 Observing magnetic interactions

STUDENT BOOK
Pages 403–7

LEARNING GOALS

Determine the properties of bulk magnetism

Explain the origins of bulk magnetism

Identify magnetic materials

Write a hypothesis

1 Materials can be classified as either ferromagnetic or non-ferromagnetic based on how they react to an external magnetic field. Outline the interaction between a ferromagnetic material and an external magnetic field.

_____

_____

_____

_____

_____

2 Katya and Sophie conducted an investigation to to classify a variety of materials as ferromagnetic or non-ferromagnetic based on the interactions they observed with an external magnetic field. They placed a $1\,cm^2$ piece of each material at a distance of $2\,cm$ directly under a permanent magnet and recorded their observations in the following table.

Material	Observation	Classification
Copper	No interaction	Non-ferromagnetic
Aluminium	No interaction	Non-ferromagnetic
Iron	Attracted towards the magnet	Ferromagnetic
Zinc	No interaction	Non-ferromagnetic
Cobalt	Attracted towards the magnet	Ferromagnetic
Nickel	Attracted towards the magnet	Ferromagnetic

Write a hypothesis for this investigation.

_____

_____

3 What are three variables that would need to be kept constant to ensure the validity of this investigation?

_____

_____

_____

4 What causes materials to have magnetic properties?

_____

_____

5 Explain in terms of their electronic structure why ferromagnetic materials react in the way they do.

_____

_____

_____

_____

_____

6 Identify three examples of ferromagnetic elements.

_____

7 Using a diagram, explain the microscopic structure of a permanent magnet.

HINT

Use domains.

_____

_____

_____

_____

As with the other field forces – gravity and the electrostatic force – we can model magnetic fields using field lines.

1 How is the direction of a magnetic field defined?

_____

_____

2 What does the relative spacing of field lines represent?

_____

_____

3 Why can field lines not cross?

_____

_____

_____

4 Draw the magnetic field lines surrounding a permanent bar magnet.

[bar magnet: S | N]

5 What would happen if the south pole of a compass was placed within the magnetic field?

_____

6 Draw the magnetic field lines for the following situations.

a

[two bar magnets: N | S   S | N]

b

7 What is happening in the region between the two north poles in question **6b**, and what does it imply?

_____

_____

An electric current also produces a magnetic field. As with any other field, it can be represented using field lines.

8 Draw the magnetic field lines surrounding the following current-carrying conductors. (Don't forget to use the right-hand grip rule.)

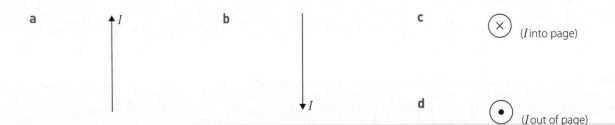

The magnetic field around a single straight current-carrying conductor is typically very weak (unless a very large current is passed through it). The magnetic field strength is often increased by coiling the conductor into a series of loops to form a solenoid.

9 Draw the magnetic field lines surrounding the following solenoids.

a

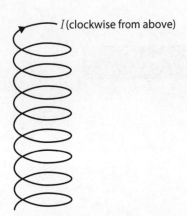

b

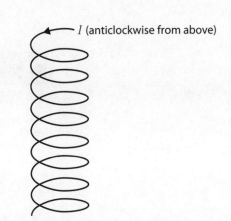

Factors affecting magnetic field strength in straight conductors

Write an experimental procedure

Create a risk assessment

Graph experimental results

Calculate constants from experimental data

Determine units

Roger and Gary conducted an investigation to determine the relationship between the current and magnetic field strength surrounding a straight conductor and the relationship between perpendicular distance from the conductor and magnetic field strength.

1 Write a suitable procedure for the first investigation.

2 Identify at least one possible risk for this experiment and how it would be mitigated.

What are the risks in doing this investigation?	How can you manage these risks to stay safe?

3 Their result averages are shown in the table below

Current (A)	Magnetic field strength (μT)
1.0	2.0
2.0	4.0
3.0	6.0
4.0	8.0
5.0	10

Using the above data, create a graph of magnetic field strength against current and insert a trend line.

4 Using the trend line that you have drawn, and the knowledge that the perpendicular distance from the wire at which the magnetic field strength was measured was 10 cm, calculate an experimental value for μ_0.

_____

_____

_____

5 Based on these results, what can you say about the accuracy of the investigation?

_____

_____

6 Roger calculated the value for μ_0 from the second investigation to be $1.6 \times 10^{-6}\,\mathrm{H\,m^{-1}}$. $(1\,\mathrm{H\,m^{-1}} = 1\,\mathrm{N\,A^{-2}})$. Explain two factors that may have caused this variation.

_____

_____

_____

_____

7 Show, using known equations, how the gradient of the above graph can be represented with units of $\mathrm{H\,m^{-1}}$.

_____

_____

_____

8 To calculate a value for μ_0 from the second investigation, Roger and Gary constructed a graph of field strength as a function of inverse distance. Why is it more appropriate to graph magnetic field strength as a function of inverse distance than as a function of distance?

_____

_____

_____

9 Explain the origin of the 2π and the need for μ_0 in the equation for the magnetic field strength of a straight conductor.

_____

_____

_____

Factors affecting magnetic field strength in solenoids

Write a hypothesis

Graph experimental results

Determine experimental relationships

Samir and Doris conducted two investigations to determine the effect of the number of coils in a solenoid on the magnetic field strength produced.

1 Write a hypothesis for this investigation.

_____

_____

The following equipment list and procedure were used to conduct the investigations.

MATERIALS

- 40m (approximately) insulated copper wire   • DC variable power supply   • 4 banana plugs   • Variable resistor
- Ammeter or multimeter   • Magnetic field probe   • 4 empty paper towel rolls   • Masking tape

METHOD

1. Cut a paper towel roll into four equal lengths of approximately 5.0cm.
2. Wind the copper wire around one of the lengths until a total of 40 coils have been produced. Ensure that the coils cover the full length of the roll and there is an even number of coils across the full length.
3. Tape each end of the coil with masking tape to prevent it from unravelling.
4. Repeat steps 2 to 3 three more times using the other previously cut lengths of paper towel roll, increasing the number of coils each time by 20 up to a maximum of 100 coils.
5. Set up a series circuit with the power supply, 40-coil solenoid, variable resistor and ammeter.
6. Set the power supply to 12V and the variable resistor to its highest setting.
7. Slowly decrease the resistance of the variable resistor until there is 1.0A flowing through the circuit.
8. Measure the magnetic field strength with the probe at the centre of the coils (if you are using a smartphone app, measure from the centre of the length on the outside at a distance of 2.0cm)
9. Repeat step 10 with the other coils and record the results in the table below.

RESULTS

The students' results table is presented below.

Number of coils	Magnetic field strength (T)
40	0.0013
60	0.0014
80	0.0024
100	0.0022

2 Create a graph of magnetic field strength as a function of number of coils.

3 Calculate the gradient of the trend line of your graph from question **2**.

_____

_____

4 Calculate an experimental value for μ_0 using the gradient calculated in question **3**. State your answer to an appropriate number of significant figures.

_____

_____

_____

5 Compare the experimental value to the accepted value of $1.3 \times 10^{-6}\,\mathrm{N\,A^{-2}}$.

_____

_____

6 Outline some limitations of this procedure that could be responsible for the discrepancy between the experimental and theoretical values for μ_0.

_____

_____

_____

_____

_____

Calculating magnetic field variables

Apply mathematical models to problems involving magnetic field variables

State your answers to an appropriate number of significant figures.

1 A long straight conductor has 2.4 A flowing through it. What is the magnetic field strength as measured from 25 cm away?

_____

2 At what distance would the magnetic field be equal to 15 μT from a conductor with 1.0 A flowing through it?

_____

3 What current would produce a magnetic field strength of 4.5 μT at a distance of 12 cm?

_____

4 The magnetic field strength around a residential power line was found to have a maximum value of 72 mT at a distance of 1.5 m. What was maximum current flowing through the line?

_____

5 A car alternator requires a minimum of 15.4 A to start the engine. What is the minimum magnetic field strength measured by a mechanic at a distance of 45 cm from a single connecting wire when the engine is started?

_____

6 A microchip has a maximum operating threshold of 200 nT. What is the maximum current that can pass through a wire that is 50 mm away from the chip before it fails?

_____

7 A solenoid is to be fitted into an 8.0 cm long void for a door-locking mechanism. How many coils must be wound to produce the 0.250 T magnetic field needed to close the lock given the maximum current drawn from the circuit is 2.40 A?

_____

8 A solenoid drawing 1.2 A of current is produced with 270 turns and has a length of 7.5 cm. What current would it need to draw to increase the magnetic field strength within the solenoid by 20%?

_____

_____

_____

9 What magnetic field strength would be produced within a solenoid that is drawing 200 mA and has a turn cm^{-1} ratio of 25?

_____

_____

10 What number of turns per centimetre for a solenoid would produce a magnetic field strength of 1.5 mT for a current of 1.75 A within the solenoid?

_____

Considering when to apply models of magnetic fields

Evaluate the use of scientific models to improve understanding

In the previous worksheets we have represented magnetic fields using both graphical (magnetic field diagram) and mathematical models. As with all scientific models, while they are great for building our understanding it is also important to consider their limitations.

1 Outline two benefits of each model of magnetic fields.

_____

_____

_____

_____

_____

_____

2 Outline two limitations of each model of magnetic fields.

_____

_____

_____

_____

_____

_____

3 Using your answers above, identify one or more criteria you would use to determine which model you would use in any given situation.

_____

_____

4 Identify two specific occasions in Stage 6 Physics in which you have applied a model to better understand a real-world physical phenomenon.

_____

_____

5 Make a judgement on the use of models in science in general based on your answers to the previous questions.

_____

_____

_____

_____

Module four: Checking understanding

Circle the correct answer for questions 1–4.

1 Which of the following is not ohmic?

   A   Copper wire

   B   Resistor component

   C   Diode

   D   Plastic

2 What resistance would be required to draw 10.0 A from a 240 V power source?

   A   2.4 Ω

   B   24 Ω

   C   0.0416 Ω

   D   0 Ω

3 What is the relationship between electric field strength produced by a point charge and the distance a test charge is from it?

   A $F \propto d$

   B $F \propto \dfrac{1}{d}$

   C $F \propto d^2$

   D $F \propto \dfrac{1}{d^2}$

4 The direction of the magnetic field is defined as:

   A   the direction in which the south pole of a compass would point.

   B   the direction in which the north pole of a compass would point.

   C   the direction in which a positive point charge would move.

   D   the direction in which a negative point charge would move.

State your answers to an appropriate number of significant figures.

5 Calculate the force acting on two helium nuclei (2 protons and 2 neutrons each) when they are placed 2.45×10^{-12} m apart.

6 a Calculate the work done accelerating an electron as it passes through a potential difference of 12 V.

   b Assuming the electron was initially stationary, calculate the final velocity of the electron.

7 Calculate the power dissipated by a kettle element with an internal electrical resistance of 27.5 Ω plugged into 240 V mains.

_____

_____

_____

8 A solenoid consisting of 275 turns of 14-gauge wire (with a diameter of 1.63 mm) is measured to be 4.25 cm long. What magnetic field strength is produced when a current of 6.10 A is drawn?

_____

_____

_____

9 Discuss the benefits and limitations of electromagnets compared to permanent magnets.

_____

_____

_____

_____

_____

10 Explain the origins of ferromagnetism.

_____

_____

_____

_____

_____

PERIODIC TABLE OF ELEMENTS

Key

Symbol of element:

	gas at room temperature
	liquid at room temperature
	solid at room temperature
	synthetic (does not occur naturally)

atomic number → 26

Fe → iron

standard atomic weight → 55.85

name of element

- s block
- p block
- d block transition metals
- d block lanthanoids and actinoids

Group	1	2	3	4	5	6	7	8	9	10	11	12	13	14	15	16	17	18
	1 **H** hydrogen 1.008																	2 **He** helium 4.003
	3 **Li** lithium 6.941	4 **Be** beryllium 9.012											5 **B** boron 10.81	6 **C** carbon 12.01	7 **N** nitrogen 14.01	8 **O** oxygen 16.00	9 **F** fluorine 19.00	10 **Ne** neon 20.18
	11 **Na** sodium 22.99	12 **Mg** magnesium 24.31											13 **Al** aluminium 26.98	14 **Si** silicon 28.09	15 **P** phosphorus 30.97	16 **S** sulfur 32.07	17 **Cl** chlorine 35.45	18 **Ar** argon 39.95
	19 **K** potassium 39.10	20 **Ca** calcium 40.08	21 **Sc** scandium 44.96	22 **Ti** titanium 47.87	23 **V** vanadium 50.94	24 **Cr** chromium 52.00	25 **Mn** manganese 54.94	26 **Fe** iron 55.85	27 **Co** cobalt 58.93	28 **Ni** nickel 58.69	29 **Cu** copper 63.55	30 **Zn** zinc 65.38	31 **Ga** gallium 69.72	32 **Ge** germanium 72.63	33 **As** arsenic 74.92	34 **Se** selenium 78.96	35 **Br** bromine 79.90	36 **Kr** krypton 83.80
	37 **Rb** rubidium 85.47	38 **Sr** strontium 87.61	39 **Y** yttrium 88.91	40 **Zr** zirconium 91.22	41 **Nb** niobium 92.91	42 **Mo** molybdenum 95.96	43 **Tc** technetium	44 **Ru** ruthenium 101.1	45 **Rh** rhodium 102.9	46 **Pd** palladium 106.4	47 **Ag** silver 107.9	48 **Cd** cadmium 112.4	49 **In** indium 114.8	50 **Sn** tin 118.7	51 **Sb** antimony 121.8	52 **Te** tellurium 127.6	53 **I** iodine 126.9	54 **Xe** xenon 131.3
	55 **Cs** caesium 132.9	56 **Ba** barium 137.3	57–71 lanthanoids	72 **Hf** hafnium 178.5	73 **Ta** tantalum 180.9	74 **W** tungsten 183.9	75 **Re** rhenium 186.2	76 **Os** osmium 190.2	77 **Ir** iridium 192.2	78 **Pt** platinum 195.1	79 **Au** gold 197.0	80 **Hg** mercury 200.6	81 **Tl** thallium 204.4	82 **Pb** lead 207.2	83 **Bi** bismuth 209.0	84 **Po** polonium	85 **At** astatine	86 **Rn** radon
	87 **Fr** francium	88 **Ra** radium	89–103 actinoids	104 **Rf** rutherfordium	105 **Db** dubnium	106 **Sg** seaborgium	107 **Bh** bohrium	108 **Hs** hassium	109 **Mt** meitnerium	110 **Ds** darmstadtium	111 **Rg** roentgenium	112 **Cn** copernicium	113 **Nh** nihonium	114 **Fl** flerovium	115 **Mc** moscovium	116 **Lv** livermorium	117 **Ts** tennessine	118 **Og** oganesson

57 **La** lanthanum 138.9	58 **Ce** cerium 140.1	59 **Pr** praseodymium 140.9	60 **Nd** neodymium 144.2	61 **Pm** promethium	62 **Sm** samarium 150.4	63 **Eu** europium 152.0	64 **Gd** gadolinium 157.3	65 **Tb** terbium 158.9	66 **Dy** dysprosium 162.5	67 **Ho** holmium 164.9	68 **Er** erbium 167.3	69 **Tm** thulium 168.9	70 **Yb** ytterbium 173.1	71 **Lu** lutetium 175.0
89 **Ac** actinium	90 **Th** thorium 232.0	91 **Pa** protactinium 231.0	92 **U** uranium 238.0	93 **Np** neptunium	94 **Pu** plutonium	95 **Am** americium	96 **Cm** curium	97 **Bk** berkelium	98 **Cf** californium	99 **Es** einsteinium	100 **Fm** fermium	101 **Md** mendelevium	102 **No** nobelium	103 **Lr** lawrencium

9780170449595

Fully worked solutions are provided below to demonstrate the steps necessary to reach the required answer. Worked solutions help you independently review your own answers.

Chapter 1: Working scientifically and depth studies

WS 1.1 PAGE 1

1 **B** is the better research question because it has a clear focus that would lead to forming a testable hypothesis. **A** is too broad a question to be able to be answered within one investigation.

2 The independent variable is the variable that is changed in the investigation, to be able to test the effects on the dependent variable.

3 The dependent variable is the variable that is measured during an investigation, to see if the change made to the independent variable has had an effect.

4 It clearly identifies the independent variable (increasing the gradient of the slope) and the dependent variable (the velocity of the ball); it makes a clear prediction that can be tested in a simple and effective way, and it is a statement.

5 It is a question, not a statement; it is not testable because it does not make any predictions; there are no clear variables in it; and it is very open-ended – what is meant by 'better'?

6 Using vulcanised rubber would make for more efficient bouncy balls than cork.

7 If your investigation does not support your hypothesis, either your hypothesis is not correct or your investigation is not testing what you think it is. You need to make sure that your experimental method is sound, and if so you can revise your hypothesis.

8 The answers can vary in this, but the general response should be that the experiment has not failed; rather, the investigators have disproved their hypothesis. In addition, the differences they observed in the way the subjects looked at the steps suggests a new research question, which could lead to formulating a hypothesis to test in the future.

9 A research question is in a question format and is used as a basis for further investigation; it is usually broad in scope. A hypothesis is a specific testable statement that is formed through research into the area.

WS 1.2 PAGE 3

1 **A** is an example of a primary (first-hand) investigation, because Jude and Natalie carried out the work themselves and used their data to formulate their answers. **B** is an example of a secondary (second-hand) investigation, as Hadja and Lachlan didn't carry out the investigations themselves but collated the information from relevant sources to formulate their own answers.

2

What are the risks in doing this investigation?	How can you manage these risks to stay safe?
Short circuit in the power supply	Check that the power supply has been electrically tested recently. Do not overload the power supply.
Light bulb may break	Ensure that the light bulbs are placed on a flat surface clear from other items.
Trip hazard over the wires	Ensure that the wires are kept clear from areas that you will walk by.

3 Answers are sample answers. As long as the student has included sensible, justifiable risks that are specific to the investigation, then their answers are correct.

What are the risks in doing this investigation?	How can you manage these risks to stay safe?
Splinters from the piece of wood	Ensure that the wood has been sanded correctly prior to use
Weights can fall onto a person	Ensure that the weights are positioned correctly on the mass carrier and that they are handled with care
Fingers may be caught in the pulley	Ensure that all hands are kept away from the pulley when it is in use

WS 1.3 PAGE 5

1 These are suggested answers only because the pros and cons will depend on the context of use.

a

Pro	Con
Accurate when measuring, no reaction time	Not always available
Can share information with others easily	Can take time to set up
Once set up easy to make multiple readings	Require careful calibration

b

Pro	Con
Nearly always available	Requires human interaction for most readings
Can download multiple apps for measuring time and distance	Expensive, so people can be hesitant to use them in situations that could cause damage
Accurate when used correctly	Can be difficult to transfer information to others from the app

c

Pro	Con
Readily available	Measures only small distances
Good for measuring small distances	Can be difficult to use when measuring vertical distances
Easy to use	Often are worn, which will affect the accuracy of the measurement reading

d

Pro	Con
Very accurate for small readings	Relies on the ability of the user to prevent error
Integrated scales so no need to use multiple measuring tools	Can be difficult to learn how to use properly
Can be used to take a variety of different readings	Not always available

2 This was not a valid investigation because the students had more than one independent variable. By changing both the launch speed and launch angle they couldn't be sure which one would have caused the change in range.

3 The students could keep the launch speed the same and only change the launch angle. They would then only have one variable being changed and all others controlled.

4 The students would need to run the experiment multiple times while getting consistent results. If their results were not consistent then they need to adjust the experiment until it can produce consistent results. If consistency cannot be achieved, they need to change their methodology.

WS 1.4 PAGE 7

1

.com.au	
Advantages	**Disadvantages**
Multiple sites available	No system of checks and balances on the information on the site
Cover a large range of topics	Often cover information on a surface level only

.gov.au	
Advantages	**Disadvantages**
Has been checked for reliability prior to being published	Possible bias due to the government policy
Often gives raw data as well as analysed data	Can use language that is hard to understand

2

Type of source	Why is it reputable?
Textbook	Has been through a thorough editing process and has been written by experts in the field
Scientific journal	Has been through a peer review process
Statements and reports from expert agencies	Is based on research that has been carried out and can be verified
White papers	While not peer reviewed, can provide data that is reliable and accurate. Make sure you check the author
Conference proceedings	Often contain early research that has not yet been peer-reviewed; as such should be used in conjunction with other sources

3

Type of source	Why is it not reputable?
Wikis and blogs	It can be edited by anyone without the reader knowing if they are experts in what they are saying
Popular media	They are biased to the viewpoint of the publisher and are not peer reviewed
Friends or teachers	The information may not be complete or they have understood it wrongly
Out-of-date research	The findings may have been superseded or shown to be wrong
Research with no references	It is likely not to have been properly researched or information that could be contradictory to the author's viewpoint may have been deliberately ignored

4 There are several different answers, for example: direct plagiarism and patchwork plagiarism. Direct plagiarism occurs where the student copies another person's work directly without attributing the work. Avoid this by rewording the source into your own words, and then quoting the original author. In patchwork plagiarism, the student copies from several different sources and then incorrectly cites or doesn't cite at all.

5

Student work	Original text (Bloggs and Bright, 2019)
When looking at the most efficient material to use when making bouncy balls, you should use vulcanised rubber as this has the greatest efficiency (Bloggs and Bright, 2019).	The investigators found that when they dropped the ball from various heights the rebound of the ball changed due to the efficiency of the material used to make the ball. They were able to identify that vulcanised rubber was the best material to use for bouncy balls.

6 Author (surname, first initial), publication date (in brackets), title, journal title, journal number, page range

WS 1.4 PAGE 9

1 Mariam had quantitative data, since it was numerical.

2 Qualitative data, since Ross used the colour change in the water to show that a reaction was taking place.

3 a Table outline not completed, no units in column headings, results not aligned correctly, no title

9780170449595

b

Results of a ball rolling down a slope	
Time (s)	Displacement (cm)
1.0	12
2.0	14
3.0	18
4.0	26
5.0	39
6.0	50
7.0	65
8.0	82
9.0	103
10.0	115

4

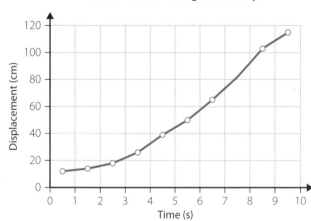

Results of a ball rolling down a slope

5 **a**

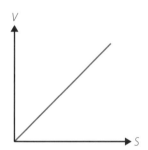

b

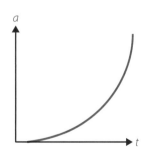

c

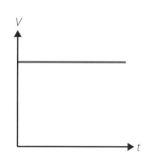

6 **a**

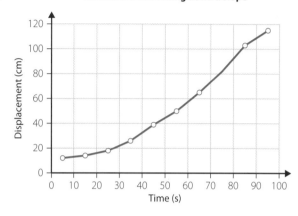

Results of a ball rolling down a slope

b From the graph, the answer should be $1.25\,\text{cm s}^{-2}$.

WS 1.5 PAGE 13

1 **a** The change in results that occurs when you are taking multiple readings. It is random and cannot be controlled for.

By taking multiple readings and working out an average. By graphing results and using the line of best fit, and by eliminating outliers from your measurements.

b Affects the results by the same amount each reading. This is generally due to an issue in the way the results are obtained or a problem in the method. These can be accounted for and generally can be mitigated by changing the equipment or method. Is predictable.

Identify the cause of the error and make sure that it is accounted for. This can include consistent calibration of the equipment as well as making sure that controls are used in the experiment and comparing results to standards.

2 Systematic: it may not be tared correctly at the start of the experiment or the zero may not be calibrated correctly.

Random: The mass being weighed may be placed in different positions when on the scale, causing it to register different masses each time.

3 The nail is $15\,\text{cm} \pm 0.5\,\text{cm}$.

4 It is between 14.5 cm and 15.5 cm.

5 $\dfrac{0.5\ \text{cm}}{15\ \text{cm}} = 0.33$

6 3%

7 **a** $P = 2(4 + 2) \pm 2(0.2 + 0.3) = 12.0 \pm 1.0\,\text{m}$

b $(4.0 - 2.0) \pm (0.2 + 0.3)\,\text{m} = 2.0 \pm 0.5\,\text{m}$

c $4.0 \times 2.0 = 8.0\,\text{m}^2$

d $\dfrac{0.2}{4.0} + \dfrac{0.3}{2.0} = 0.2$

e As there is a high % error of 20% the absolute error of the area would be $1.6\,\text{m}^2$.

8 **B** would support the hypothesis because the results would fall into this range with the uncertainty.

9 Because the satellite requires several readings to be able to triangulate the position of the receiver, it is important for the readings for Katy's location to be very precise. The readings are also required to be accurate because this will give the receiver the best results. If the results were accurate but not precise then the receiver would only know Katy's general area and not her precise location, which would not be very helpful. If the information is precise but not accurate then it would be able to specify an exact location but this may not be where Katy actually is. Therefore, it is important to be both accurate and precise or the information given would not be correct.

WS 1.6 PAGE 16

1 The improvement in technology allows for more accurate predictions or testing of the model.

2 That an atom was mostly empty space with a small, solid, positively charged core, rather than solid with dispersed charges as in the the plum pudding model described by Thomson. This was because experimental evidence showed little or no

~continued in right column ▲

deflection of most of the positive alpha particles fired at the atom but strong deflection of a small number of alpha particles.

3 The planetary model proposed by Rutherford could not account for the stability of electron orbits and why electrons did not lose energy and fall into the nucleus. Therefore, something else must be happening to explain why atoms remain stable.

4 a Mathematical modelling using computers, because this will allow for known information to be input into the system to give the possible outcomes on a large scale.

b Mathematical modelling, because this would help to show the relationship of expansion of the universe with distance. For example, Hubble's model for the expanding universe shows a linear relationship.

c Physical modelling, because the investigators would be able to see how the materials involved in the crash would behave in the real world.

WS 1.7 PAGE 17

1

	Video	PowerPoint or oral presentation	Written report
Pros	Good if the presenter gets nervous public speaking. Can include special effects if required. Can include all the required information without being interrupted. Can be viewed when time permits	Can interact with the audience to get the point across. Can use props or models for the audience to interact with and gain understanding.	Good practice for scientific writing. Able to take time to formulate your response to ensure the question has been answered
Cons	Audience is unable to ask the presenter questions. Cannot use props for the audience. Unable to justify information if viewed away from the presenter	Often rely on the information in the slides and end up reading it as opposed to presenting it. Can be information overload. Time consuming to do correctly.	Sometimes difficult to put understanding into words. Time consuming to research and write properly.

2 a The abstract is a brief description of what was studied. It details the general findings and allows for the reader to decide the paper's purpose and if the paper is relevant to their interests.

b The purpose of the introduction is to show what will be addressed in the paper. This will include the prior research that the investigator has undertaken, the inquiry question and the hypothesis.

c Method **B** would be the correct one as this is very clear and shows the steps required to complete the investigation concisely. Method **A** is too general and its language is not in the correct scientific terminology.

d They should be displayed in the form they were recorded, without being changed or analysed first. This should include any outliers or unexpected results. They should be what was seen and not what was hoped to have been seen.

e It is important to analyse your results to see if they have any trends in the data. This will allow you to be able to say if your hypothesis has been supported or not.

f **A** is the more suitable answer since the investigators showed that they can see a way they can improve their experiment without getting more equipment which may not be available. It also shows a more realistic understanding of their investigation.

g The answer would be **A** as the conclusion is linking the results back to the hypothesis. In **B** the results are not related back to the hypothesis so the results don't actually answer it.

h If you have received assistance with collecting your data or in your analysis this is where you say so. It is also a place to say thank you for other assistance such as the loan of specialist equipment.

i A reference list shows all the sources that were cited within the report, whereas a bibliography shows all the sources that the information used in a report was gathered from. This includes information that was not directly included in the report.

MODULE ONE: KINEMATICS

REVIEWING PRIOR KNOWLEDGE PAGE 20

1 a We use gradient $= \frac{\text{rise}}{\text{run}}$: rise $= 3 - 1 = 2$, run $= 2 - 0 = 2$.

So the gradient is $\frac{2}{2} = 1$. The units are the units of y (m s^{-1}) divided by the units of x (s). Gradient is 1 m s^{-2}.

b We use $\frac{\text{rise}}{\text{run}}$: rise $= 0 - 3 = -3$, run $= 5 - 3 = 2$.

So the gradient is $\frac{-3}{2} = -1.5 \text{ m s}^{-2}$.

c The area under the line between 0 and 2 consists of a rectangle and a triangle:

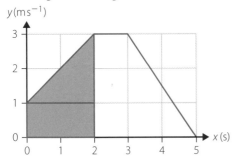

The area of the rectangle is $1 \times 2 = 2$, the area of the triangle is $\frac{1}{2} \times 2 \times 2 = 2$. So the total area is 4. The units are the units of x (s) multiplied by the units of y (m s^{-1}), so area is 4 m. We can get the same answer by counting squares off the graph, though this does not work as well when the slopes are not as convenient as 45°.

d Using the same ideas as above, we get an area of 10 m.

2 Yes it can. The area under the graph below is negative because the area between the graph and the x-axis is greater below than above.

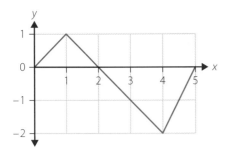

3 a Yes – acceleration has a direction, in this case *down*.

  b No – volume does not have a direction. It has a size only.

  c Yes – velocity has a direction, in this case 'towards my friend'.

4 $v^2 - u^2 = 2as$

$$\frac{v^2 - u^2}{2s} = a$$

5 a c is the hypotenuse and the known side is adjacent to the angle, so we use cosine:

$$\cos 28° = \frac{4.2 \text{ cm}}{c}$$
$$c = \frac{4.2 \text{ cm}}{\cos 28°}$$
$$= 4.7 \text{ cm}$$

  b b is opposite the angle and the known side is adjacent, so we use tan.

$$\tan 28° = \frac{b}{4.2 \text{ cm}}$$
$$b = 4.2 \text{ cm} \times \tan 28°$$
$$= 2.1 \text{ cm}$$

  c
$$a^2 + b^2 = c^2$$
$$(4.2 \text{ cm})^2 + (2.1 \text{ cm})^2 = 22.05 \text{ cm}^2$$
$$\sqrt{22.05 \text{ cm}^2} = 4.7 \text{ cm} = c$$

Chapter 2: Motion in a straight line

WS 2.1 PAGE 22

1 A scalar has a magnitude only; it does not have direction. For example, age (e.g. 15 years) is a scalar.

A vector has both magnitude and direction. For example, displacement (e.g. 100 km north) is a vector.

2

Quantity	Scalar or vector?	SI unit
time	scalar	s
distance	scalar	m
displacement	vector	m
speed	scalar	m s^{-1}
velocity	vector	m s^{-1}
acceleration	vector	m s^{-2}

3 Sam's maximum possible displacement is 36 km, if he travels the same direction all the time in a straight line.

Sam's minimum possible displacement is 0 km, if he takes a path that leads him back to where he started.

4 Speed is how fast something is moving; velocity is how fast and in what direction. Speed is distance divided by time; velocity is displacement divided by time.

5 Velocity and acceleration do not have to be in the same direction. Whenever you are slowing down your velocity and acceleration are in opposite directions.

6 $\vec{v} = \dfrac{\vec{s}}{t}$

The velocity is the rate of change of the displacement. Both are vectors and they point in the same direction.

WS 2.2 PAGE 24

1 One possible hypothesis: That instantaneous velocity will vary with time, between zero and some maximum positive and negative values. Many other answers are possible.

2

What are the risks in doing this investigation?	How can you manage these risks to stay safe?
e.g. The object may fly off the spring and hit someone	Attach the object firmly, use a soft object, wear safety glasses

3, 4, 6

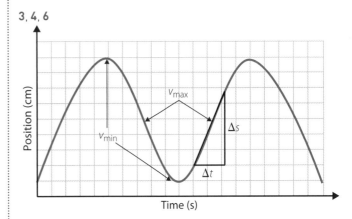

5 The value of the minimum instantaneous velocity is zero, which occurs at the top and bottom of the curve.

6 Draw a tangent to the graph at a point of maximum velocity.

See graph. The gradient is the rise over the run for this line, $\vec{v} = \text{gradient} = \dfrac{\Delta \vec{s}}{\Delta t}$. Make sure you include units as you go, so your units at the end are correct!

7 Any time period in which the position is not the same at the end as at the beginning will do. Measure the length of the time period, Δt, from the graph and the change in position, $\Delta \vec{s}$, and the average velocity is then $\vec{v}_{\text{ave}} = \dfrac{\Delta \vec{s}}{\Delta t}$.

8 Any time period in which the object returns to its starting position will give a zero average velocity. It does not matter where in the motion you choose to start the time period, as long as the objects returns to the same position.

9 a The instantaneous velocity varies sinusoidally with time – between zero when the object is at the top and bottom of its path, and having maximum positive and negative values in between.

b The average velocity is zero over a very long period of time, or over any time period in which the object ends where it began. For short time periods, the average velocity approaches the instantaneous velocity.

c How you can improve the experiment depends on how you carried it out. You could, for example, check for additional variables that need to be controlled other than those you are intentionally varying. You could make repeat measurements to check reliability. Measuring over a larger number of oscillations reduces uncertainty in average velocity.

10 This will depend on your hypothesis and the results of your investigation.

WS 2.3 **PAGE 27**

1 A spatial coordinate system that allows you to measure positions from a specified origin, and hence measure velocities and accelerations.

2 The ground, or Earth's surface.

3

Velocities of A and B	$\vec{v}_A - \vec{v}_B$ positive or negative?	A and B getting closer together or further apart?
$\vec{v}_A$ positive and $\vec{v}_B$ negative, position of A is positive and position of B is negative	Positive	Further apart
$\vec{v}_A$ negative and $\vec{v}_B$ positive, position of A is positive and position of B is negative	Negative	Closer together
$\vec{v}_A$ and $\vec{v}_B$ both positive and equal	Zero	Neither
$\vec{v}_A$ and $\vec{v}_B$ both positive, $\vec{v}_A$ greater than $\vec{v}_B$ and A has the larger displacement from the reference point	Positive	Further apart
$\vec{v}_A$ and $\vec{v}_B$ both positive, $\vec{v}_A$ greater than $\vec{v}_B$ and B has the larger displacement from the reference point	Positive	Closer together

4 a negative, positive
b positive, negative
c negative, positive
d positive, negative

5 The speed relative to you of a car going in the same direction as you is smaller than the speeds shown on its speedometer. For a car coming towards you, the relative speed is greater. This is because speedometers show the speed relative to the ground. The relative speed of cars to each other is the difference in their velocities – if the velocities have the opposite sign the relative speed is the sum of the two speeds. If the velocities have the same sign, the relative speed is the difference in the two speeds.

6 a $\vec{v}_{\text{car}} - \vec{v}_{\text{Olga}} = 60\,\text{km h}^{-1} - 40\,\text{km h}^{-1} = 20\,\text{km h}^{-1}$
b $\vec{v}_{\text{Olga}} - \vec{v}_{\text{car}} = 40\,\text{km h}^{-1} - 60\,\text{km h}^{-1} = -20\,\text{km h}^{-1}$
c No, it makes no difference.

WS 2.4 **PAGE 29**

1 $5\,\text{minutes} = 5\,\text{min} \times \dfrac{60\,\text{s}}{\text{min}} = 300\,\text{s}$
$v = \dfrac{\Delta d}{\Delta t} = \dfrac{200\,\text{m}}{300\,\text{s}} = 0.67\,\text{m s}^{-1}$

2 Lei starts with 0 initial velocity, so $\Delta v = 8.6\,\text{m s}^{-1}$.

$\vec{a} = \dfrac{\Delta \vec{v}}{\Delta t} = \dfrac{8.6\,\text{m s}^{-1}}{2\,\text{s}} = 4.3\,\text{m s}^{-2}$

3 a Distance: $s = 200\,\text{m}$

b $v = \dfrac{s}{\Delta t} = \dfrac{200\,\text{m}}{28\,\text{s}} = 7.1\,\text{m s}^{-1}$

c $\vec{s} = 0\,\text{m}$ because Lei returns to the same position.

d $\vec{v} = \dfrac{\vec{s}}{\Delta t} = 0\,\text{m s}^{-1}$

4 a $v = \dfrac{s}{\Delta t}$

So $\Delta t = \dfrac{s}{v} = \dfrac{500\,\text{m}}{1.4\,\text{m s}^{-1}} = 357\,\text{s}$, which is approximately 6 minutes.

b $2\,\text{hours} = 2\,\text{h} \times \dfrac{60\,\text{min}}{\text{h}} \times \dfrac{60\,\text{s}}{\text{min}} = 7200\,\text{s}$
$v = \dfrac{s}{\Delta t}$
So $s = v\Delta t = 1.4\,\text{m s}^{-1} \times 7200\,\text{s} = 10\,080\,\text{m}$, which is approximately 10 km.
Elena walks very briskly!

5 $100\,\text{km h}^{-1} = 100\,\text{km h}^{-1} \times \dfrac{1000\,\text{m}}{1\,\text{km}} \times \dfrac{1\,\text{h}}{3600\,\text{s}} = 28\,\text{m s}^{-1}$
The car ends at zero speed, so $\Delta v = 28\,\text{m s}^{-1}$.
$\vec{a} = \dfrac{\Delta \vec{v}}{\Delta t}$

So $\Delta t = \dfrac{\Delta \vec{v}}{\vec{a}} = \dfrac{28\,\text{m s}^{-1}}{11\,\text{m s}^{-2}} = 2.5\,\text{s}$

6 a 9 km – read directly from the graph. 9 km = 9000 m.
b Read the time from the graph: $t = 50\,\text{minutes} = 3000\,\text{s}$.

$v = \dfrac{s}{\Delta t} = \dfrac{9000\,\text{m}}{3000\,\text{s}} = 3\,\text{m s}^{-1}$

9780170449595

c

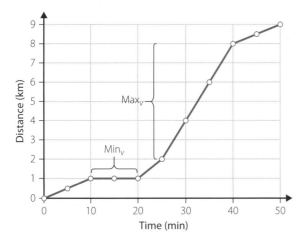

d For the region shown, the displacement and time read from the graph are 6 km = 6000 m and 15 min = 900 s.

$$v = \frac{\Delta d}{\Delta t} = \frac{6000 \text{ m}}{900 \text{ s}} = 6.7 \text{ m s}^{-1}$$

7

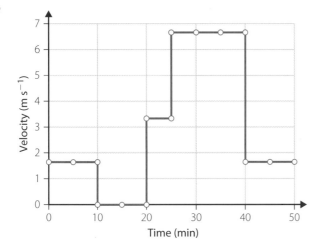

8 a The maximum acceleration is in the first 2 minutes when the plane goes from 0 km h^{-1} to 500 km h^{-1}.
2 min = 120 s and

$$500 \text{ km h}^{-1} = 500 \text{ km h}^{-1} \times \frac{1000 \text{ m}}{1 \text{ km}} \times \frac{1 \text{ h}}{3600 \text{ s}} = 139 \text{ m s}^{-1}$$

$$\vec{a} = \frac{\Delta \vec{v}}{\Delta t} = \frac{139 \text{ m s}^{-1}}{120 \text{ s}} = 1.2 \text{ m s}^{-2}$$

b The total displacement is the area under the graph. We can estimate this by counting rectangles, and then multiplying by the area of each rectangle. There are approximately equal numbers of partial rectangles that are smaller than a half rectangle and that are larger than a half rectangle. So we count the total number of partial rectangles, divide by two, and add that to the number of complete rectangles.
Each rectangle has a height of $100 \text{ km h}^{-1} = 28 \text{ m s}^{-1}$ and a width of 5 min = 300 s.
So the area of each rectangle is height × width
$= 28 \text{ m s}^{-1} \times 300 \text{ s} = 8400 \text{ m}$ or 8.4 km.
There are 74 complete rectangles under the curve and 22 partial rectangles, which we approximate to 85, so the total distance is $85 \times 8.4 \text{ km} = 710 \text{ km}$. This is rounded to two significant figures because the method is imprecise – 700 km is a reasonable estimate.

c $v = \frac{\Delta d}{\Delta t} = \frac{710 \text{ km}}{1 \text{ h}} = 710 \text{ km h}^{-1}$

$$v = \frac{\Delta d}{\Delta t} = \frac{710\,000 \text{ m}}{3600 \text{ s}} = 200 \text{ m s}^{-1} \text{ rounded to two significant}$$

figures, because the method is imprecise.

WS 2.5 PAGE 32

1 Substitute $\Delta \vec{v} = \vec{v} - \vec{u}$ into $\vec{a} = \frac{\Delta \vec{v}}{\Delta t}$: $\vec{a} = \frac{\Delta \vec{v}}{\Delta t} = \frac{\vec{v} - \vec{u}}{\Delta t}$

then rearrange for $\vec{v}$: $\vec{v} = \vec{u} + \vec{a}t$

2 a $\Delta t = \frac{\Delta \vec{v}}{\vec{a}}$

b $\Delta t = \frac{\vec{v} - \vec{u}}{\vec{a}}$

c $\vec{s} = \frac{(\vec{v} + \vec{u})}{2} \frac{(\vec{v} - \vec{u})}{\vec{a}}$

d $\vec{s} = \frac{(\vec{v} + \vec{u})}{2} \frac{(\vec{v} - \vec{u})}{\vec{a}} = \frac{\vec{v}^2 - \vec{u}^2}{2\vec{a}}$

$$\vec{v}^2 - \vec{u}^2 = 2\vec{a}\vec{s}$$
$$\vec{v}^2 = \vec{u}^2 + 2\vec{a}\vec{s}$$

3 a $\vec{s} = \vec{u}t + \frac{1}{2}\vec{a}t^2$. In this case $u = 0$, so we can simplify the equation to $\vec{s} = \frac{1}{2}\vec{a}t^2$.

Rearrange for t: $t = \sqrt{\frac{2\vec{s}}{\vec{a}}} = \sqrt{\frac{2(2.0 \text{ m})}{9.8 \text{ m s}^{-2}}} = 0.64 \text{ s}$

b Using the equation for t derived above, and the lower value of g for the Moon:

$$t = \sqrt{\frac{2\vec{s}}{\vec{a}}} = \sqrt{\frac{2(2.0 \text{ m})}{1.6 \text{ m s}^{-2}}} = 1.6 \text{ s}$$

c $\vec{s} = \vec{u}t + \frac{1}{2}\vec{a}t^2 = 0 + \frac{1}{2}(9.8 \text{ m s}^{-2})(1.6 \text{ s})^2 = 7.8 \text{ m}.$

4 Here $v = 100 \text{ km h}^{-1} = 28 \text{ m s}^{-1}$ and $u = 60 \text{ km h}^{-1} = 17 \text{ m s}^{-1}$.

$$\vec{v}^2 = \vec{u}^2 + 2\vec{a}\vec{s}$$

$$\vec{a} = \frac{\vec{v}^2 - \vec{u}^2}{2s} = \frac{(28 \text{ m s}^{-1})^2 - (17 \text{ m s}^{-1})^2}{2(200 \text{ m})} = 1.2 \text{ m s}^{-2}$$

5 Here $u = 100 \text{ km h}^{-1} = 28 \text{ m s}^{-1}$ and $v = 110 \text{ km h}^{-1} = 30.5 \text{ m s}^{-1}$.

$\vec{v} = \vec{u} + \vec{a}t$
Rearrange:

$$t = \frac{\vec{v} - \vec{u}}{\vec{a}} = \frac{30.5 \text{ m s}^{-1} - 28 \text{ m s}^{-1}}{1.2 \text{ m s}^{-2}} = 2.1 \text{ s}$$

6 Now $u = 100 \text{ km h}^{-1} = 28 \text{ m s}^{-1}$ and $v = 60 \text{ km h}^{-1} = 17 \text{ m s}^{-1}$.

We can use either $\vec{s} = \vec{u}t + \frac{1}{2}\vec{a}t^2$ or $\vec{v}^2 = \vec{u}^2 + 2\vec{a}\vec{s}$ but for either we first need the acceleration.

Assuming uniform acceleration,

$\vec{a} = \frac{\Delta \vec{v}}{\Delta t} = \frac{\vec{v} - \vec{u}}{\Delta t} = \frac{17 \text{ m s}^{-1} - 28 \text{ m s}^{-1}}{6 \text{ s}} = -1.8 \text{ m s}^{-2}.$ Note that
we keep an extra decimal place and round at the final step.

Now, using $\vec{s} = \vec{u}t + \frac{1}{2}\vec{a}t^2 = (28 \text{ m s}^{-1})(6 \text{ s})^2 = 136 \text{ m}.$ (Note that
if we keep more significant figures at intermediate steps we get 135 m.)

WS 3.1 PAGE 34

1 No. We get the length of the vector from the magnitudes of its two orthogonal (perpendicular) components using Pythagoras's theorem. That equation cannot give a result smaller than either of the numbers that goes into it.

2 Yes. If one of the components has zero magnitude, then the vector will have a magnitude equal to that of the other component.

3 a Length from Pythagoras, $|\vec{a}| = \sqrt{|\vec{a}_x|^2 + |\vec{a}_y|^2}$. If we say that a is the length of $\vec{a}$, then

$$a = \sqrt{a_x^2 + a_y^2}$$

$$a_y = \sqrt{a^2 - a_x^2} = \sqrt{(35\,\text{cm})^2 - (28\,\text{cm})^2} = 21\,\text{cm}$$

b If we assume that the components point in the positive axis directions, we get:

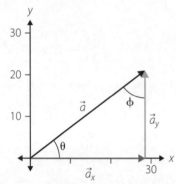

c $\tan\theta = \dfrac{a_y}{a_x}$

$$\theta = \tan^{-1}\left(\frac{a_y}{a_x}\right) = \tan^{-1}\left(\frac{21\,\text{cm}}{28\,\text{cm}}\right) = \tan^{-1}(0.75) = 37°$$

$$\phi = \tan^{-1}\left(\frac{a_x}{a_y}\right) = \tan^{-1}\left(\frac{28\,\text{cm}}{21\,\text{cm}}\right) = \tan^{-1}(1.33) = 53°$$

And we can see that $53° + 37° = 90°$ as we would expect.

4

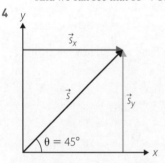

$\tan\theta = \dfrac{\text{Opposite}}{\text{Adjacent}} = \dfrac{s_y}{s_x}$ and if the angle is 45° then, because $\tan 45° = 1$, we can say $s_x = s_y$.

Now $s^2 = s_x^2 + s_y^2 = 2s_x^2$ so that $s_x = \dfrac{s}{\sqrt{2}} = \dfrac{15}{\sqrt{2}} = 10.6\,\text{cm}$

$\approx 11\,\text{cm}$ (2 significant figures) in the x direction.

And s_y has the same magnitude, and is in the y direction.

Alternatively, we can use trigonometry:

$s_x = 15\sin 45° = 10.6\,\text{cm}$ in the x direction.

$s_y = 15\cos 45° = 106\,\text{cm}$ in the y direction

5 a b

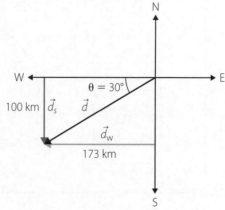

c See above. We use trigonometry, recalling that

$$\sin\theta = \frac{\text{Opposite}}{\text{Hypotenuse}} \quad \text{and} \quad \cos\theta = \frac{\text{Adjacent}}{\text{Hypotenuse}}.$$

As we have drawn it, $\vec{d}_S$ is opposite and $\vec{d}_W$ is adjacent.

$d_S = d\sin\theta = (200\,\text{km})\sin 30° = 100\,\text{km south}$

$d_W = d\cos\theta = (200\,\text{km})\cos 30° = 173\,\text{km west}$

Note that we can use Pythagoras to check our answer.

6 a

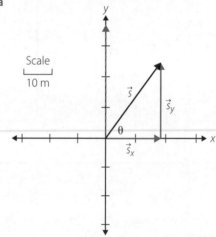

b We know that the components given are mutually perpendicular, so we can use Pythagoras to get the length of the vector:

$$s = \sqrt{s_x^2 + s_y^2} = \sqrt{(15\,\text{m})^2 + (20\,\text{m})^2} = 25\,\text{m}$$

c To make sure we define our angle correctly, we used our diagram, which shows that $\vec{s}_y$ is opposite and $\vec{s}_x$ is adjacent.

$$\tan\theta = \frac{\text{Opposite}}{\text{Adjacent}} = \frac{s_y}{s_x}$$

$$\theta = \tan^{-1}\left(\frac{s_y}{s_x}\right) = \tan^{-1}\left(\frac{20\,\text{m}}{15\,\text{m}}\right) = \tan^{-1}(1.33) = 53°$$

Note: After part **b**, we had all three side lengths. We could use the cosine rule, or the sine rule, to work out θ. For example we can use the sine rule: $\dfrac{A}{\sin A} = \dfrac{B}{\sin B} = \dfrac{C}{\sin C}$, and we know that the angle opposite the hypotenuse is 90°.

So:

$$\frac{s}{\sin 90°} = \frac{s_y}{\sin\theta}$$

9780170449595

Recall that $\sin 90° = 1$, so we can rearrange as:

$$\sin\theta = \frac{s_y}{s}$$

$$\theta = \sin^{-1}\left(\frac{s_y}{s}\right)$$

$$= \sin^{-1}\left(\frac{20\,\text{m}}{25\,\text{m}}\right)$$

$$= \sin^{-1}(0.8) = 53°$$

One way to decide on a method is to ask: Does one of these methods avoid the use of intermediate results? The first method above only uses numbers given in the question. The sine rule method uses a result we derived (the hypotenuse) and so may give a less accurate and precise answer. Thus, the first method is preferable.

7 Begin with a sketch, to make sure we define θ correctly.

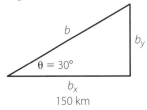

We can see that b_y is opposite the angle θ and b_x is adjacent.

$$\tan\theta = \frac{\text{Opposite}}{\text{Adjacent}} = \frac{b_y}{b_x}$$

We then rearrange to isolate the things of interest:

$$b_y = b_x \tan\theta = (150\,\text{km})\tan 30° = 86.6\,\text{km} = 87\,\text{km}$$

$$b_x = b\cos\theta$$

$$b = \frac{b_x}{\cos\theta} = \frac{150\,\text{km}}{\cos 30°} = 173\,\text{km}$$

We can check using Pythagoras's theorem, the sine and/or the cosine rule.

8 Take $\vec{v}$ as the vector and v as its magnitude. We must take care with our trigonometry to use sine and cosine correctly – it depends on which angle we are given. SOH–CAH–TOA is always handy. The sine rule may also be useful, to find or check answers.

a $v_x = v\cos 60° = 15\cos 60°\,\text{m} = 7.5\,\text{m}$

$v_y = v\sin 60° = 15\sin 60°\,\text{m} = 13\,\text{m}$

We could use the sine rule: $\dfrac{A}{\sin A} = \dfrac{B}{\sin B} = \dfrac{C}{\sin C}$

$$\frac{v}{\sin 90°} = \frac{v_y}{\sin 60°} = \frac{v_x}{\sin 30°}$$

or, recalling $\sin 90° = 1$:

$v_y = v\sin 60° = 15\sin 60°\,\text{m} = 13\,\text{m}$

$v_x = v\cos 60° = 15\cos 60°\,\text{m} = 7.5\,\text{m}$

The remaining answers use only one method, but feel free to choose another.

b $v_y = \sqrt{v^2 - v_x^2} = \sqrt{(15\,\text{m})^2 - (11\,\text{m})^2}$

$\quad = 10.2\,\text{m} = 10\,\text{m}$ to 2 signficant figures

c Note that because the angle is given to the vertical, not the horizontal, we use sin to get the x component and cos to get the y.

$v_x = v\sin 40° = (12\,\text{m})\sin 40° = 7.7\,\text{m}$

$v_y = v\cos 40° = (12\,\text{m})\cos 40° = 9.2\,\text{m}$

d $v = \sqrt{v_y^2 + v_x^2} = \sqrt{(12\,\text{m})^2 + (9\,\text{m})^2} = 15\,\text{m}$

(You may recognise this one as a 3–4–5 triangle.)

WS 3.2 PAGE 37

1 a, c

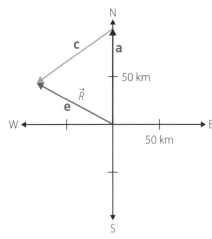

b $\vec{s} = 100\,\text{km N}$

d Distance is not a vector, so we can add up without thinking about the vector nature of the quantities. Distance is like 'what does my odometer/trip-meter say?'

$100\,\text{km} + 100\,\text{km} = 200\,\text{km}$

e Use a ruler and note the scale and work out how long the vector is. You should use a protractor to get the angle. A sensible answer might be:

75 km, 20° north of west.

If worked out mathematically, the answer is 76.5 km at an angle 22.5° north of west. You may discuss the relative merits of the two methods – mathematics is more precise, measurement may be more intuitive.

f Distance is greater. Because displacement is the *vector* addition of the two legs of the journey, part of one component can cancel out part of the other.

2 b

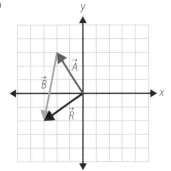

c

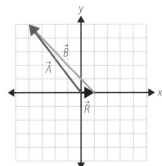

3 a

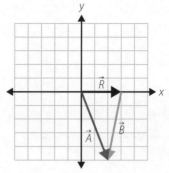

b

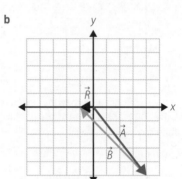

4 a A diagram is useful.

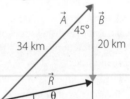

The only eastward (E) component comes from $\vec{A}$, and because it is at a 45° angle, we know that its easterly and northerly components are:

$\vec{A}_E = (34\,km)\cos 45° = 24\,km$ E and

$\vec{A}_N = (34\,km)\sin 45° = 24\,km$ N

$\vec{B}$ is 20 km southward, so subtract it from the northward component of $\vec{A}$, leaving 4 km N.

Thus, the resulting displacement is the vector sum of 24 km E and 4 km N.

To get the vector direction, we could use these components. However, they are derived results. If possible, we should refer back to the original, given quantities. If doing so, we write:

$$\tan\theta = \frac{(34\,\cos 45° - 20)\,km}{34\,\sin 45°\,km}$$

so that $\theta = 9.5°$

If written in terms of intermediate results, we would write:

$\tan\theta = \dfrac{4\,km}{24\,km}$ so that $\theta = \tan^{-1}\left(\dfrac{4\,km}{24\,km}\right) = 9.5°$

When working in this way, keep lots of decimal places, and only round at the very end when quoting a result.

Similarly, R, the length of $\vec{R}$, is:

$R = \sqrt{(34\,\cos 45° - 20)^2 + (34\,\sin 45°)^2}$ km

$\quad = 24.4$ km

We could have used:

$R = \sqrt{(4\,km)^2 + (24\,km)^2} = 24.4$ km

but that would be less precise.

Regardless, we find that $\vec{R}$ is 24.4 km long and is at an angle of 9.5° north of east.

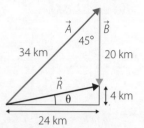

b This will just be the negative of $\vec{R}$; a vector 24.4 km long and 9.5° south of west.

5 Draw our compass points and draw in paths $\vec{A}$ and $\vec{B}$, approximately to scale. Convert all measurements to the same units – metres in this case.

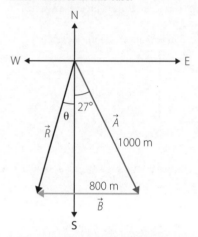

We recall our trigonometry – SOH–CAH–TOA. The length of the southward component of $\vec{A}$, A_S, is given by $A_S = (1000\,m)\cos 27° = 891\,m$.

The eastward component is $A_E = (1000\,m)\sin 27° = 454\,m$.

$\vec{B}$ runs west, so the total westward displacement is $800\,m - 454\,m = 346\,m$ west.

Resultant final displacement magnitude in terms of the given quantities is: $R = \sqrt{(1000\,\cos 27°)^2 + (800 - 1000\,\sin 27°)^2}$ m

$\quad = 955.83\,m = 956\,m$

To get the angle west of south,

$$\tan\theta = \frac{(800 - 1000\,\sin 27°)\,m}{1000\,\cos 27°\,m}$$

$\theta = 21.2°$

Students can use $R = \sqrt{(891)^2 + (346)^2}$ m $= 956\,m$ and

$\tan\theta = \dfrac{346\,m}{891\,m}$, but should keep all possible decimal places

through the intermediate steps.

The resultant is: 956 m S21.2°W.

6 a To get to where Julie is from where Bryan is, we go 550 m W (the negative horizontal direction) and 150 m S (also negative), so relative to Bryan Julie is at (−550 m, −150 m).

b

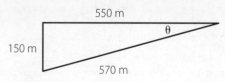

From Pythagoras, the length of the resultant vector is

$R = \sqrt{(-550\,m)^2 + (-150\,m)^2} = 570\,m$.

The side opposite θ is 150 m, and adjacent is 550 m, so the direction is:

$\tan\theta = \dfrac{150\,m}{550\,m}$, so that $\theta = 15.3°$.

We must make sure we know what our definition of θ is. In this case, Julie is 570 m from Bryan in a direction 15.3° south of west.

9780170449595

7 a If the total vector length is S (for Sandeep) and the components east and north are S_E and S_N, then we have to choose whether to use sine or cosine based on how the angle is defined. It is relative to the vertical (N) so we use sine for the easterly component.

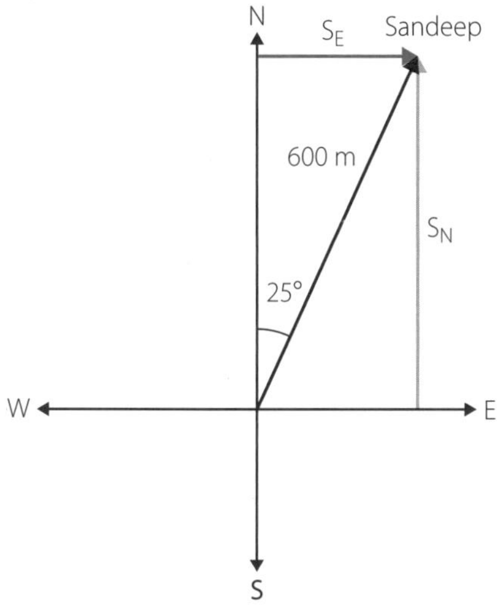

~continued in right column

$S_E = S\sin\theta = (600\,m)\ \sin 25° = 254\,m\ E$

b $S_N = S\cos\theta = (600\,m)\cos 25° = 544\,m\ N$

c First, how far west of the shop is Małgosia?

$M_W = M\sin\theta = (500\,m)\ \sin 40° = 321\,m\ W$

So Małgosia is 321 m + 254 m = 575 m west of Sandeep.

Ideally, to avoid rounding errors we would not use intermediate results. Thus, it would be better to write

$(600\,m)\sin 25° + (500\,m)\sin 40° = 574.96\,m = 575\,m$

In this case, the two methods give the same result, but we could imagine a case in which rounding before adding might give a sightly wrong answer. On the other hand, expressions can get very complicated if we do not use intermediate results. As a useful compromise, always keep intermediate results to more decimal places than you will quote in your final result.

d We do a vector sum of the north–south components for Sandeep and Małgosia:

$(500\,m)\ \cos 40° + (600\,m)\cos 25° = 927\,m\ S$

e Avoiding rounding, we find:

$$\text{Total distance} = \sqrt{((500\text{ m})\ \cos 40° + (600\text{ m})\ \cos 25°)^2 + ((600\text{ m})\ \sin 25° + (500\text{ m})\ \sin 40°)^2}$$
$$= 1.09\text{ km}$$

We can see from this that we cannot always expect to write our answers in terms of the original information. Expressions become unwieldy. It is reasonable to write:

$$\text{Total distance} = \sqrt{(926.807\text{ m})^2 + (574.965\text{ m})^2}$$
$$= 1090.67\text{ m} = 1.09\text{ km}$$

WS 3.3 PAGE 41

1 a Distance is a scalar. Think about what a car odometer would read.

Total distance = 100 km + 120 km = 220 km

b Average speed $= \dfrac{\text{total distance}}{\text{total time}}$

$$t = \frac{d}{v} = \frac{100\text{ km}}{100\text{ km h}^{-1}} + \frac{120\text{ km}}{80\text{ km h}^{-1}} = 2.5\text{ h}$$

$$\text{Average speed} = \frac{220\text{ km}}{2.5\text{ h}} = 88\text{ km h}^{-1}$$

c We begin by drawing a scale diagram. This helps us define our angles and directions. Draw in the legs of the journey, as if adding vectors by putting them nose to tail. The final, or resultant, displacement is labelled $\vec{R}$.

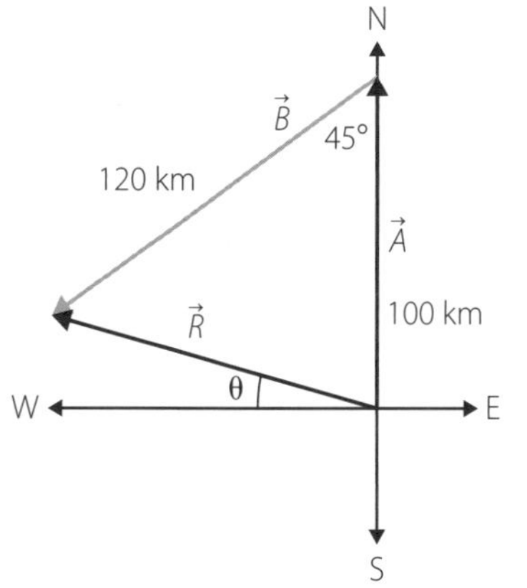

The total westward distance comes from $\vec{B}$, and we might say:

$B_W = B\sin 45° = (120\,km)\sin 45° = 84.4\,km$

$B_S = B\cos 45° = (120\,km)\cos 45° = 84.4\,km$

where B_W is the magnitude of the westerly component of $\vec{B}$, for example.

Total north–south displacement is:

$100\,km\ N + (120\,km)(\cos 45°)S$

$= 100\,km - (120\,km)(\cos 45°)N$

So resultant displacement length is

$$R = \sqrt{((120\text{ km})\sin 45°)^2 + (100\text{ km} - (120\text{ km})\cos 45°)^2}$$
$$= 86.194\text{ km} = 86.2\text{ km}.$$

Angle to the W direction is:

$$\tan\theta = \frac{\text{Opposite}}{\text{Adjacent}} = \frac{100\text{ km} - (120\text{ km})\cos 45°}{(120\text{ km})\sin 45°}$$
$$\theta = 10.1°$$

Average velocity is therefore

$$\frac{\text{resultant displacement}}{\text{total time}} = \frac{86.194\text{ km}}{2.5\text{ h}}$$
$$= 34.5\text{ km h}^{-1}\ \text{W}10.15°\text{N}.$$

d Speed is larger because when calculating the displacement some of the distances cancel out. It cannot be the other way around.

They would be the same if the two legs of the trip were parallel (going the same way).

2 a We build on answers from worksheet 3.2 question **4a**. We can obtain the displacement from Worksheet 3.2, question **4a**. We find that the resultant vector is:

$$\vec{R} = 24.4\,\text{km E9.5°N}$$

Allen has 20 minutes $= \dfrac{1}{3}$ hour to get back. If we take R as the magnitude of the displacement, then:

$$v = \frac{R}{t} = \frac{24.4\ \text{km}}{\frac{1}{3}\ \text{h}} = 73\,\text{km h}^{-1}$$

b Let's call the unknown time t in hours. So total time is now $(1 + t)$ hours.

$$\text{average speed} = \frac{\text{distance}}{\text{total time}}$$
$$= \frac{(34 + 20 + 24.4)\ \text{km}}{(1 + t)\ \text{h}} = 50\,\text{km h}^{-1}$$

We rearrange and find t.

$$(34 + 20 + 24.4)\,\text{km} = (50\,\text{km h}^{-1})(1 + t)\,\text{h}$$

$$t = \frac{(34 + 20 + 24.4 - 50)\ \text{km}}{50\ \text{km h}^{-1}} = 0.568\,\text{h} = 34\ \text{minutes}$$

3 a For speed, we can just add the distances and the times. No need to worry about working with vectors. We can probably do that without a diagram! Recall, 15 minutes is 0.25 h and 45 minutes is 0.75 h. We use $d = vt$ to get the distances of the legs.

$$\text{Average speed} = \frac{\text{total distance}}{\text{total time}}$$
$$= \frac{\text{distance}_1 + \text{distance}_2}{\text{time}_1 + \text{time}_2}$$
$$= \frac{(0.25\ \text{h} \times 12\ \text{km h}^{-1} + 0.75\ \text{h} \times 6.0\ \text{km h}^{-1})}{(0.25 + 0.75)\ \text{h}}$$
$$= 7.5\,\text{km h}^{-1}$$

b

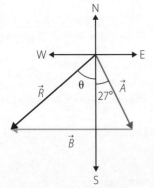

We need to find the total displacement, R, the magnitude of $\vec{R}$. The southward part all comes from $\vec{A}$ (of length $A = 3.0\,\text{km}$).

$$R_{\text{S}} = A_{\text{S}} = A\cos 27° = (3.0\,\text{km})\cos 27° = 2.673\,02\,\text{km}$$

In this example, we are keeping extra decimal places, which we'll round off at the end.

The westward part comes from both segments. Length of $\vec{B}$ is 4.5 km.

$$R_{\text{W}} = B_{\text{W}} + A_{\text{W}} = 4.5\,\text{km} - (3.0\,\text{km})\sin 27° = 3.128\,03\,\text{km}$$
$$R = \sqrt{(2.673\,02\,\text{km})^2 + (3.138\,03\,\text{km})^2} = 4.122\,17\,\text{km}$$
$$= 4.1\,\text{km}$$

$$\tan\theta = \frac{3.138\,03\,\text{km}}{2.673\,02\,\text{km}} \text{ so that } \theta = 49.6°$$

$$\text{average velocity} = \frac{\text{displacement}}{\text{total time}} = \frac{4.12217\ \text{km}}{1.0\ \text{h}}$$
$$= 4.1\,\text{km h}^{-1}\ \text{S49.6°E}.$$

4 a The ball started with $8.0\,\text{m s}^{-1}$ one way, and finished with $6.0\,\text{m s}^{-1}$ the opposite way. A change of $-8.0\,\text{m s}^{-1}$ brings it to zero, and another $-6.0\,\text{m s}^{-1}$ sends it back the other way. Mathematically:

$$\Delta\vec{v} = \vec{v}_{\text{f}} - \vec{v}_{\text{i}} \text{ (i = initial, f = final)}$$
$$\Delta\vec{v} = -6.0\,\text{m s}^{-1} - 8.0\,\text{m s}^{-1} = -14\,\text{m s}^{-1} \text{ towards the wall}$$
$$= 14\,\text{m s}^{-1} \text{ away from the wall.}$$

b $\vec{a} = \dfrac{\Delta\vec{v}}{\Delta t} = \dfrac{-14\ \text{m s}^{-1}}{2\ \text{s}} = -7\,\text{m s}^{-2}$ towards the wall

$$= 7\,\text{m s}^{-2} \text{ away from the wall.}$$

c Less, because the real time of the impact will be much less than 2 s. To cause the same change of velocity in a shorter time, the acceleration must be greater.

Also, even within the real time of the impact, the acceleration will not be uniform. It will be small when the ball first touches the wall, for example, and has not yet deformed very much, and will peak for a very brief time near the centre of the collision.

d We know initial and final velocities, and these are constant with time. We have a narrow transition in between, where the bounce occurs.

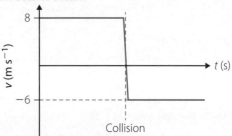

e Acceleration is the gradient of velocity with time. The graph in part **d** has zero gradient in most places and a strong negative gradient in the bounce region.

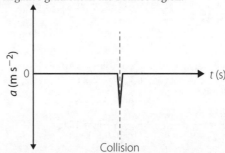

5 a There is no change, because the direction of the velocity has changed but the speed has not.

b At C we have $12\,\text{m s}^{-1}$ to the right $= -12\,\text{m s}^{-1}$ to the left. (To add the velocities, we put both in the same coordinate system.)

$$\Delta\vec{v} = \vec{v}_{\text{f}} - \vec{v}_{\text{i}} \text{ (i = initial, f = final, where } A \text{ is initial and } C \text{ is final)}$$
$$\Delta\vec{v} = (-12 - 12)\,\text{m s}^{-1} \text{ to the left} = -24\,\text{m s}^{-1} \text{ to the left}$$
$$= 24\,\text{m s}^{-1} \text{ to the right}$$

c There is no change, because the direction of the velocity has changed but the speed has not.

d At A we have $12\,\text{m s}^{-1}$ to the left.

At B we have $12\,\text{m s}^{-1}$ to the rear.

$$\Delta\vec{v} = \vec{v}_{\text{f}} - \vec{v}_{\text{i}} \text{ (i = initial, f = final, where } A \text{ is initial and } B \text{ is final)}$$
$$\Delta\vec{v} = \vec{v}_B - \vec{v}_A$$

Let us draw the vectors, then recall that

$$\Delta\vec{v} = \vec{v}_B - \vec{v}_A = \vec{v}_B + (-\vec{v}_A)$$

9780170449595

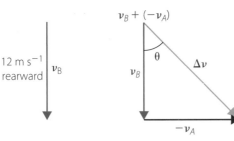

We can see that the magnitude of $\Delta \vec{v}$ (Δv) comes from Pythagoras:

$$\Delta v = \sqrt{(12\ \text{m s}^{-1})^2 + (12\ \text{m s}^{-1})^2} = 17\ \text{m s}^{-1}$$

And because the two components are of equal magnitude, we can see that $\Delta \vec{v}$ will point in a direction back and to the right (negative left) at an angle of 45°.

6 a From the equations in chapter 2, we know:

$$v_\parallel = u_\parallel + at$$

where $\parallel$ means 'parallel with' (down) the slope. Note that the initial velocity parallel with the slope is $u_\parallel = 0$, so we can say $v_\parallel = at$.

b It is constant, because there is no source of acceleration or friction in this direction. Thus, $v_\perp = v_\text{i}$.

c We begin by drawing a vector diagram. At any time, the puck will have a speed down the slope ($\parallel$) and across the slope ($\perp$). The total velocity of the puck, $\vec{v}$, is the vector sum of these two. Its magnitude is written v.

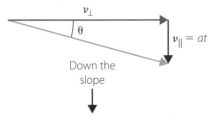

So from Pythagoras (again!) we can write:

$$v = \sqrt{v_\parallel{}^2 + v_\perp{}^2} = \sqrt{(at)^2 + v_\text{i}{}^2}$$

because $v_\parallel = at$ and $v_\perp = v_\text{i}$.

d $\tan\theta = \dfrac{\text{Opposite}}{\text{Adjacent}} = \dfrac{v_\parallel}{v_\perp} = \dfrac{at}{v_\text{i}}$ or $\theta = \tan^{-1}\left(\dfrac{at}{v_\text{i}}\right)$

e When t is small, $\dfrac{at}{v_\text{i}}$ will be small, the angle will be small and the puck will be moving in the initial direction, perpendicular to the slope. When t is big, $\dfrac{at}{v_\text{i}}$ will be big, and θ will get closer and closer to 90°. That is, early on the puck is going across the slope but later it is mostly going down the slope. That makes sense because its 'down the slope' velocity is increasing but its 'across the slope' velocity is not.

1

Describe a situation in which …	Description
The displacement of one object relative to the other is constant	They have the same velocity (they are moving in parallel directions and keeping the same speed; $v = 0$ for both is one case).
The displacement of one object relative to the other is increasing at a constant rate (that is, linearly)	Two objects moving at different velocities (one may be zero)
The displacement of one object relative to the other is increasing at an increasing rate	Two objects with differing accelerations, moving away from each other
The objects have velocities opposite in direction and the relative displacement is decreasing	Objects heading towards each other (not necessarily to collide)

2 a We use $\vec{s}_{1\ \text{relative to}\ 2} = \vec{d}_1 - \vec{d}_2$ or $\vec{s}_{\text{K relative to M}} = \vec{d}_\text{K} - \vec{d}_\text{M}$
Away from the wall is the positive direction, so we can say:
$10 \times 2.5\ \text{m} - 4 \times 2.5\ \text{m} = 15\ \text{m}$

b It will be the reverse, so -15 m, or -6 spaces.

3 The first leg of each journey is the same for both. Because we are working out a relative displacement, we can ignore it and we leave it off our diagram. We take east and north as positive. The vector denoted $\vec{R}$ is the one that we want.

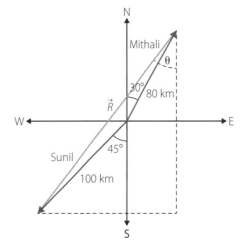

Mithali (M) is at coordinates $((80\ \text{km})\sin 30°\text{E}, (80\ \text{km})\cos 30°\text{N})$
Sunil (S) is at $((-100\ \text{km})\sin 45°\text{E}, (-100\ \text{km})\cos 45°\text{N})$
We do the subtraction:
$\vec{s}_{\text{S rel to M}} = \vec{s}_\text{S} - \vec{s}_\text{M}$
$= ((-100\ \text{km})\sin 45°\text{E}, (-100\ \text{km})\cos 45°\text{N})$
$\quad - ((80\ \text{km})\sin 30°)\text{E}, (80\ \text{km})\cos 30°\text{N})$
$= (-100\ \text{km})\sin 45° - (80\ \text{km})\sin 30°$ east and
$\quad (-100\ \text{km})\cos 45° - (80\ \text{km})\cos 30°$ north
$= -124.85\ \text{km}$ east and $-140.00\ \text{km}$ north
$= 125\ \text{km}$ west and $140\ \text{km}$ south

The length of the displacement is $\sqrt{124.85^2 + 140.00^2} = 188\ \text{km}$ (to 3 significant figures, so we kept 5 significant figures in the intermediate results).

The angle of the vector is $\tan\theta = \dfrac{124.85}{140.00}$ or $\theta = 42°$.

So Sunil is 188 km S42°W of Mithali.

4 The two components here are perpendicular. If we take the velocity of the ground as zero, then we have

$$\vec{v}_{\text{Y relative to G}} = \vec{v}_Y - \vec{v}_G = \vec{v}_Y$$

The yoyo is going up the string and the string is going forward as Lauren runs. We combine the two velocities, making sure the units are the same.

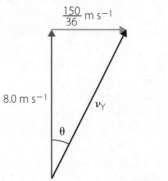

$$15\,\text{km h}^{-1} = 15\,000\,\text{m h}^{-1} = \frac{15000\,\text{m}}{3600\,\text{s}} = \frac{150}{36}\,\text{m s}^{-1}\ (\approx 4.2\,\text{m s}^{-1})$$

It is easy to use the given values, so we do not use an intermediate value for the walking speed in m s^{-1}. Note that 8.0 m s^{-1} is given to 2 significant figures.

$$v_{\text{Y relative to G}} = \sqrt{\left(8.0\,\text{m s}^{-1}\right)^2 + \left(\frac{150}{36}\,\text{m s}^{-1}\right)^2} = 9.0\,\text{m s}^{-1}$$

The angle is $\tan\theta = \dfrac{\left(\dfrac{150}{36}\,\text{m s}^{-1}\right)}{8.0\,\text{m s}^{-1}}$ or $\theta = 27.5°$

where the direction of the vector is upwards and forwards at an angle of about 28° to the vertical.

5 a The two components here are perpendicular. If we take the velocity of the ground as zero, then we have

$$\vec{v}_{\text{Y relative to G}} = \vec{v}_Y - \vec{v}_G = \vec{v}_Y$$

Lauren is running east, The yoyo is swinging north at A. We combine the two velocities, making sure the units are the same.

A diagram helps:

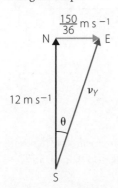

Magnitude from Pythagoras:

$$v_{\text{Y relative to G}} = \sqrt{\left(12.0\,\text{m s}^{-1}\right)^2 + \left(\frac{150}{36}\,\text{m s}^{-1}\right)^2} = 12.7\,\text{m s}^{-1}$$

The angle is $\tan\theta = \dfrac{\left(\dfrac{150}{36}\,\text{m s}^{-1}\right)}{12.0\,\text{m s}^{-1}}$ or $\theta = 19°$.

We can express this as N19°E.

b There is no change in speed. The yoyo at both places is moving perpendicular to the running direction, so the net speed will be unchanged.

c At both points the yoyo has an easterly velocity of 15.0 km h^{-1}, so this can be ignored when finding a change rather than an absolute value. At A the yoyo has a left-to-right velocity of -12 m s^{-1}. At C has a left-to-right velocity of 12 m s^{-1}. So the change is 24 m s^{-1} in a left to right direction (that is, north).

d From part **a**, we know the speed at A is 12.7 m s^{-1}. At B, it is going forwards because of the running but backwards because of the swinging, and its net speed is:

$$v_B = 12.0\,\text{m s}^{-1}\,(\text{backwards}) + \frac{150}{36}\,\text{m s}^{-1}\,(\text{forwards})$$
$$= 7.8\,\text{m s}^{-1}\,(\text{backwards})$$
$$\Delta v = v_A - v_B = (12.7 - 7.8)\,\text{m s}^{-1} = 4.9\,\text{m s}^{-1}$$

Because we are working with a speed, we do not have to do vector subtraction or quote a direction.

e At both points the yoyo has an easterly velocity of 15.0 km h^{-1}, so this can be ignored when finding a change rather than an absolute value. At A the yoyo has a northerly velocity due to the swing of 12 m s^{-1}. At B it has a easterly velocity from the swing of -12 m s^{-1} (negative because it is going west) relative to Lauren.

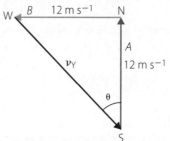

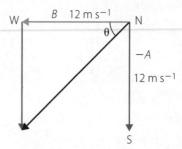

$$\vec{v}_{\text{at }B\text{ relative to at }A} = \vec{v}_{\text{at }B} - \vec{v}_{\text{at }A} = \vec{v}_{\text{at }B} + (-\vec{v}_{\text{at }A})$$
$$= (0\,\text{m s}^{-1}\,\text{north} - 12\,\text{m s}^{-1}\,\text{east}) - (12\,\text{m s}^{-1}\,\text{N} + 0\,\text{m s}^{-1}\,\text{E}).$$

The magnitude of the change is:

$$\Delta v = \sqrt{(-12.0\,\text{m s}^{-1})^2 + (-12.0\,\text{m s}^{-1})^2} = 17.0\,\text{m s}^{-1}$$

And because the components are equal, we can see the direction will be 45° south–west.

6 a The easterly components come from both the running and the easterly component of the motion at E. The southerly component comes from the motion at E. So we can add the two motions as vectors.

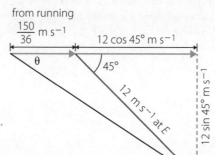

9780170449595

At E we have an eastward component from the swing of $12\,\mathrm{m\,s^{-1}}\cos 45°$ and a southward component of $12\,\mathrm{m\,s^{-1}}$ $\sin 45°$. We also have a forward component from the running of $\dfrac{150}{36}\,\mathrm{m\,s^{-1}}$. Thus the speed is the Pythagorean sum of the two components:

$$v_{\mathrm{East}} = (12\ \mathrm{m\,s^{-1}})\cos 45° + \frac{150}{36}\ \mathrm{m\,s^{-1}}$$

$$v_{\mathrm{south}} = (12\ \mathrm{m\,s^{-1}})\sin 45°$$

$$v_{\mathrm{Y\ relative\ to\ G}} = \sqrt{(v_{\mathrm{east}})^2 + (v_{\mathrm{south}})^2}$$

$v_{\mathrm{Y\ relative\ to\ G}}$

$$= \sqrt{\left((12\ \mathrm{m\,s^{-1}})\cos 45° + \frac{150}{36}\ \mathrm{m\,s^{-1}}\right)^2 + \left((12\ \mathrm{m\,s^{-1}})\sin 45°\right)^2}$$

$$= 15.2\ \mathrm{m\,s^{-1}}$$

The direction is to the south of east. We can see that v_{south} is opposite the angle, so

$$\tan\theta = \frac{(12\ \mathrm{m\,s^{-1}})\sin 45°}{(12\ \mathrm{m\,s^{-1}})\cos 45° + \frac{150}{36}\ \mathrm{m\,s^{-1}}}\ \text{ or }\ \theta = 34°$$

So:

$\vec{v} = 15\,\mathrm{m\,s^{-1}}$ 34° south of east.

b At F, the situation is the mirror of that at E – there is now a northward instead of a southward component, but otherwise it is the same. So we can say:

$\vec{v} = 15\,\mathrm{m\,s^{-1}}$ 34° north of east.

Symmetry is often a very powerful way of coming up with answers.

c The forward component is unchanged. We go from pointing south to pointing north, so the direction of the velocity change will be to the north.

At E the southward component was of size $12\,\mathrm{m\,s^{-1}}\sin 45°$, so at F we have a northward component of the same size, so the total change in velocity will be $2 \times 12\,\mathrm{m\,s^{-1}}\sin 45°$ $= 17\,\mathrm{m\,s^{-1}}$ north.

WS 3.5 **PAGE 48**

1 a $\vec{v}_W = \vec{v}_A - \vec{v}_W = 700\,\mathrm{km\,h^{-1}} - 200\,\mathrm{km\,h^{-1}}$
$= 500\,\mathrm{km\,h^{-1}}$ west

b Treat the velocity to the east as a negative one to the west:
$\vec{v}_{A\ \mathrm{rel\ to\ W}} = \vec{v}_A - \vec{v}_W = 700\,\mathrm{km\,h^{-1}} - (-200\,\mathrm{km\,h^{-1}})$
$= 900\,\mathrm{km\,h^{-1}}$ west

c If the wind is blowing at $200\,\mathrm{km\,h^{-1}}$ to the north but the aeroplane is maintaining a westerly course relative to the ground, the aeroplane must have a $200\,\mathrm{km\,h^{-1}}$ velocity component that is southerly.

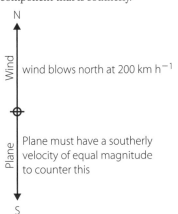

wind blows north at 200 km h⁻¹

Plane must have a southerly velocity of equal magnitude to counter this

d i The plane has a $200\,\mathrm{km\,h^{-1}}$ velocity component to the south. At the same time, the plane is moving at $700\,\mathrm{km\,h^{-1}}$ west, perpendicular to the wind.

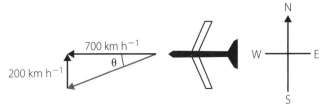

$$v = \sqrt{(700\ \mathrm{km\,h^{-1}})^2 + (200\ \mathrm{km\,h^{-1}})^2} = 728\ \mathrm{km\,h^{-1}}$$

ii We can use trigonometry to get the angle. From the diagram in part **i**, we can see that the side opposite to θ is $200\,\mathrm{km\,h^{-1}}$ and the adjacent is $700\,\mathrm{km\,h^{-1}}$.

$$\tan\theta = \frac{\mathrm{Opposite}}{\mathrm{Adjacent}} = \frac{200\ \mathrm{km\,h^{-1}}}{700\ \mathrm{km\,h^{-1}}}\ \text{ or }\ \theta = 16°$$

So the pilot must aim the plane in a direction 16° south of west and travel at an airspeed of $728\,\mathrm{km\,h^{-1}}$ to actually travel west at $700\,\mathrm{km\,h^{-1}}$ relative to the ground.

1 Plane aims at 728 km h⁻¹ 16° south-west.

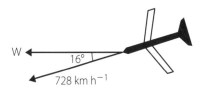

2 This gives it a 200 km h⁻¹ southerly speed and a 700 km h⁻¹ westerly speed.

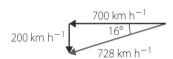

3 The wind blows north at 200 km h⁻¹. This cancels out the southerly part of the plane's speed, leaving it travelling at 700 km h⁻¹ directly west.

2 a We have a forward speed of $0.50\,\mathrm{m\,s^{-1}}$ and a left-to-right speed of $0.80\,\mathrm{m\,s^{-1}}$. Pythagoras gives the speed, v, as:

$$v = \sqrt{(0.80\ \mathrm{m\,s^{-1}})^2 + (0.50\ \mathrm{m\,s^{-1}})^2} = 0.94\ \mathrm{m\,s^{-1}}$$

b The diagram in the question shows that the current is opposite θ and the rowing speed is adjacent, so:

$$\tan\theta = \frac{\mathrm{Opposite}}{\mathrm{Adjacent}} = \frac{0.80\ \mathrm{m\,s^{-1}}}{0.50\ \mathrm{m\,s^{-1}}}\ \text{ or }\ \theta = 58°$$

c We use $v = \dfrac{d}{t}$ where v is velocity, d is distance and t is time. Rearrange to get $t = \dfrac{d}{v}$:

$$t = \frac{50\ \mathrm{m}}{0.50\ \mathrm{m\,s^{-1}}} = 100\ \mathrm{s}$$

Time to go $50\,\mathrm{m}$ at $0.50\,\mathrm{m\,s^{-1}}$ is $100\,\mathrm{s}$. We then rearrange the same formula to get $d = vt$, but now the velocity is the current:

$$d = vt = 0.80\,\mathrm{m\,s^{-1}} \times 100\,\mathrm{s} = 80\,\mathrm{m}$$

An $0.80\,\mathrm{m\,s^{-1}}$ current will carry Diana 80 m downstream in that time.

d The component of her rowing must counter the current, so her upstream component must be equal in magnitude and opposite in direction to the current – that is, 0.80 m s^{-1} to the left. She must also have some component perpendicular to the bank, or she will never get across!

e

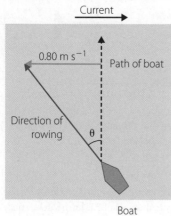

f We use $v = \dfrac{d}{t}$ where v is velocity, d is distance and t is time.

$$v = \frac{d}{t} = \frac{50.0 \text{ m}}{50.0 \text{ s}} = 1.00 \text{ m s}^{-1}$$

50.0 m in 50.0 s requires 1.00 m s^{-1}

g

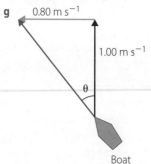

We can get the magnitude of the velocity from Pythagoras:

$$v = \sqrt{(0.80 \text{ m s}^{-1})^2 + (1.00 \text{ m s}^{-1})^2} = 1.28 \text{ m s}^{-1}$$

And we can see that 0.80 m s^{-1} is opposite θ and 1.00 m s^{-1} is adjacent, so:

$$\tan\theta = \frac{\text{Opposite}}{\text{Adjacent}} = \frac{0.80 \text{ m s}^{-1}}{1.00 \text{ m s}^{-1}} \text{ or } \theta = 38.7°$$

where the angle is defined as away from the perpendicular to the bank. We could also define the angle as 'to the current', in which case it would be 51.3°.

3 a Note that the diagram below illustrates the velocities, not the positions, so all the arrows have their tails on the origin, even though the cars may all be in different places.

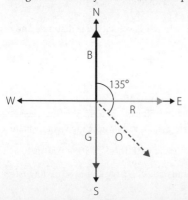

b The green car and the black one are going in exactly opposite directions. Therefore we can work with the magnitudes – we don't need to do vector maths. B = black, G = green. v_G is negative because we take north, N, as positive. So we have
$\vec{v}_B = 100 \text{ km h}^{-1} \text{ N}$ and $\vec{v}_G = -80 \text{ km h}^{-1} \text{ N}$
$\vec{v}_{G \text{ relative to B}} = \vec{v}_G - \vec{v}_B = (-80 \text{ km h}^{-1} - 100 \text{ km h}^{-1}) \text{ N}$
$= -180 \text{ km h}^{-1} \text{ N} = 180 \text{ km h}^{-1} \text{ S}$

So the cars are moving away from each other at a relative speed of 180 km h^{-1}, which is what we might expect.

c The red (R) car is not moving in a parallel direction with the black car, so we must use vectors. First, we write each velocity as a vector sum of northward and eastward components:
$\vec{v}_B = 100 \text{ km h}^{-1} \text{ N} + 0 \text{ km h}^{-1} \text{ E}$ and
$\vec{v}_R = 0 \text{ km h}^{-1} \text{ N} + 80 \text{ km h}^{-1} \text{ E}$
Then, subtract the northward components and, separately, the eastward ones:
$\vec{v}_{R \text{ relative to B}} = \vec{v}_R - \vec{v}_B$
$= (0 \text{ km h}^{-1}) - 100 \text{ km h}^{-1})$
$\qquad \text{N} + (80 \text{ km h}^{-1} - 0 \text{ km h}^{-1}) \text{ E}$
$= -100 \text{ km h}^{-1} \text{ N} + 80 \text{ km h}^{-1} \text{ E}$
Use Pythagoras to get the magnitude of the relative velocity:
$$\sqrt{(-100 \text{ km h}^{-1})^2 + (80 \text{ km h}^{-1})^2} = 128 \text{ km h}^{-1}$$
Now, we might want to draw a diagram to make sure we find θ correctly.

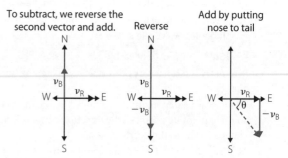

The magnitude of the side opposite θ is 100 km h^{-1}, and the adjacent is 80 km h^{-1}, so:

Direction: $\tan\theta = \dfrac{\text{Opposite}}{\text{Adjacent}} = \dfrac{100 \text{ km h}^{-1}}{80 \text{ km h}^{-1}}$ or

$\theta = 51.3°$ south of east

d Up to now, the cars have been moving either parallel or perpendicular to each other. Here, we consider a more general case. For consistency, we measure all angles as away from the N direction. O = orange. As before, we write each velocity as a vector sum of northerly and easterly components:
$\vec{v}_B = 100 \text{ km h}^{-1} \text{ N} + 0 \text{ km h}^{-1} \text{ E}$ and
$\vec{v}_O = 80 \sin(-135°) \text{ km h}^{-1} \text{ N} + 80 \cos(-135°) \text{ km h}^{-1} \text{ E}$
Then we do the vector subtraction to get the relative velocity:
$\vec{v}_{O \text{ relative to B}} = \vec{v}_O - \vec{v}_B$

$= (80 \cos(-135°) \text{ km h}^{-1} - 100 \text{ km h}^{-1}) \text{ N}$
$+ (80 \sin(-135°) \text{ km h}^{-1} - 0 \text{ km h}^{-1}) \text{ E}$

The magnitude of the relative velocity is then

$$v_{\text{O relative to B}} = \sqrt{\left(80\cos(-135°)\ \text{km h}^{-1} - 100\ \text{km h}^{-1}\right)^2 + \left(80\sin(-135°)\ \text{km h}^{-1} - 0\ \text{km h}^{-1}\right)^2} = 166\ \text{km h}^{-1}$$

Direction: $\tan\theta = \dfrac{80\sin(-135°)\ \text{km h}^{-1}}{80\cos(-135°)\ \text{km h}^{-1} - 100\ \text{km h}^{-1}}$ or $\theta = 20°$ east of south

Here is a vector diagram. The easterly component of v_{O} has magnitude $80\sin(-135°)\,\text{km h}^{-1}$ and that gives the opposite side of the triangle, while the length of v_{B} and the southerly component of v_{O} ($80\cos(-135°)\,\text{km h}^{-1}$) give the adjacent:

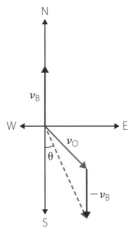

It may be simpler in a case like this to work out the components, note them down to several decimal places, then do the remaining calculations and only round off when quoting final results.

e This is a difficult question and is considered an extension. The idea is to use it as an example of generating a formula that can answer many different cases, rather than doing a separate calculation for each case. Use Y for yellow.

We measure the angle as away from the N direction, like we did in part **d** – we just do not substitute in numbers at the end:

$\vec{v}_{\text{B}} = 100\ \text{km h}^{-1}\ \text{N} + 0\ \text{km h}^{-1}\ \text{E}$ and
$\vec{v}_{\text{Y}} = 80\sin\theta\ \text{km h}^{-1}\ \text{N} + 80\cos\theta\ \text{km h}^{-1}\ \text{E}$

$\vec{v}_{\text{Y relative to B}} = \vec{v}_{\text{Y}} - \vec{v}_{\text{B}}$

$\qquad = (80\sin\theta\ \text{km h}^{-1} - 100\ \text{km h}^{-1})\ \text{N}$

$\qquad + (80\cos\theta\ \text{km h}^{-1} - 0\ \text{km h}^{-1})\ \text{E}$

The magnitude of the relative velocity is then

$v_{\text{Y relative to B}}$

$= \sqrt{(80\cos\theta\ \text{km h}^{-1} - 100\ \text{km h}^{-1})^2 + (80\sin\theta\ \text{km h}^{-1})^2}$

The curve looks like this:

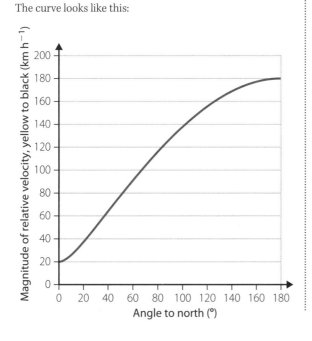

Let's consider the cases we've already done.

Green car: $\theta = 180°$. Formula gives $180\ \text{km h}^{-1}$ (we can see this from the graph).

Red car: $\theta = 90°$. Formula gives $128\ \text{km h}^{-1}$.

Orange car: $\theta = 135°$. Formula gives $166\ \text{km h}^{-1}$.

And we can see that the graph makes sense for $\theta = 0°$, too, where we simply get the difference between the two speeds: $100 - 80 = 20\ \text{km h}^{-1}$.

4 a

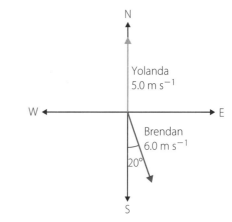

b $\vec{v}_{\text{Y rel to B}} = \vec{v}_{\text{Y}} - \vec{v}_{\text{B}} = \vec{v}_{\text{Y}} + (-\vec{v}_{\text{B}})$

$\vec{v}_{\text{Y}} = 0\ \text{m s}^{-1}\ \text{E} + 5.0\ \text{m s}^{-1}\ \text{N}$

$\vec{v}_{\text{B}} = (6.0\sin20°)\ \text{m s}^{-1}\ \text{E} + (-6.0\cos20°)\ \text{m s}^{-1}\ \text{N}$

$\vec{v}_{\text{Y rel to B}} = [0\ \text{m s}^{-1} - (6.0\sin20°)\ \text{m s}^{-1}]\ \text{E}$

$\qquad\qquad + [5.0\ \text{m s}^{-1} - (-6.0\cos20°)\ \text{m s}^{-1}]\ \text{N}$

$\qquad = -2.1\ \text{m s}^{-1}\ \text{E} + 10.6\ \text{m s}^{-1}\ \text{N}$

$\qquad = 2.1\ \text{m s}^{-1}\ \text{W} + 10.6\ \text{m s}^{-1}\ \text{N}$

If we wish to maintain precision, we can substitute values in later.

$v_{\text{Y relative to B}}$
$$= \sqrt{(-6.0\sin20°\,\text{km h}^{-1})^2 + (5.0\,\text{km h}^{-1} + 6.0\cos20°\,\text{m s}^{-1})^2}$$
$$= 10.8\,\text{m s}^{-1}$$

$$\tan\theta = \frac{-6.0\sin20°\,\text{m s}^{-1}}{5.0\,\text{m s}^{-1} + 6.0\cos20°\,\text{m s}^{-1}} \text{ or } \theta = 11°$$

So relative to Brendan, Yolanda is moving at $10.8\,\text{m s}^{-1}$ N11°W.

c

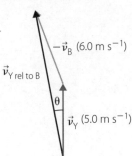

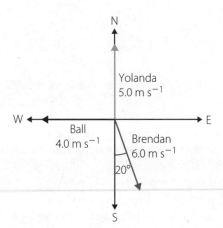

We must compare $v_{\text{Y rel to ball}} = \vec{v}_\text{Y} - \vec{v}_\text{ball}$ and $v_{\text{B rel to ball}} = \vec{v}_\text{B} - \vec{v}_\text{ball}$.

So the question is, which vector is longer, $\vec{v}_\text{Y} + (-\vec{v}_\text{ball})$ or $\vec{v}_\text{B} + (-\vec{v}_\text{ball})$? This is a qualitative question, so we can use a scale vector diagram to show us it must be Brendan.

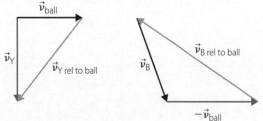

MODULE ONE: CHECKING UNDERSTANDING PAGE 52

1 **B** Displacement
2 **D** By working out the area under the graph.
3 **A** 1.04 s
4 **C** 1000 m
5 **C**
6 **A** $14.8\,\text{m s}^{-1}$

Note: We have an easterly velocity of $60.0\,\text{km h}^{-1} = 13.9\,\text{m s}^{-1}$; we have a left-to-right (southerly) component of $v = \dfrac{d}{t} = 5.00\,\text{m s}^{-1}$. Pythagoras gives the total magnitude.

7 First, write our data out using established notation.
$u = 4.00\,\text{km s}^{-1}$, $v = 11.0\,\text{km s}^{-1}$, $s = 100\,\text{km}$ and $a = $ unknown. Which equation links these four quantities?
$v^2 = u^2 + 2as$

Rearrange to make a the subject
$$a = \frac{v^2 - u^2}{2s} = \frac{(11.0\,\text{km s}^{-1})^2 - (4.00\,\text{km s}^{-1})^2}{2(100\,\text{km})}$$

because we are using km s^{-1} as our unit of velocity and km for distance, it is useful to leave the units in to make sure they cancel out correctly.
$a = 0.725\,\text{km s}^{-2} = 725\,\text{m s}^{-2}$

8 a

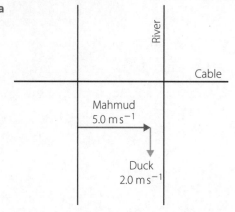

b From the chapter, we have:
$$\vec{v}_{1\,\text{relative to 2}} = \vec{v}_1 - \vec{v}_2$$
or
$$\vec{v}_{\text{M relative to D}} = \vec{v}_\text{M} - \vec{v}_\text{D}$$
Now, $\vec{v}_\text{M} - \vec{v}_\text{D} = \vec{v}_\text{M} + (-\vec{v}_\text{D})$
So, let us sketch that:

We get the magnitude from Pythagoras:
$$\left|\vec{v}_{\text{M relative to D}}\right| = v_{\text{M relative to D}} = \sqrt{(v_\text{M})^2 + (v_\text{D})^2}$$
$$= \sqrt{(5.0\,\text{m s}^{-1})^2 + (2.0\,\text{m s}^{-1})^2}$$
$$= \sqrt{29\,(\text{m s}^{-1})^2}$$
$$= 5.4\,\text{m s}^{-1}$$

As for the direction, we can use trigonometry:
$$\tan\theta = \frac{2.0\,\text{m s}^{-1}}{5.0\,\text{m s}^{-1}} = 0.4$$
$$\theta = \tan^{-1}(0.4) = 22°$$

So $\vec{v}_{\text{M relative to D}}$ is $5.4\,\text{m s}^{-1}$ at an angle 22° upstream of perpendicular to the bank.

MODULE TWO: DYNAMICS

REVIEWING PRIOR KNOWLEDGE PAGE 55

1 a Balanced – velocity is constant
  b Unbalanced – velocity is changing
  c Balanced – velocity is constant
  d Unbalanced – velocity is changing
  e Unbalanced – velocity is changing because direction changes

9780170449595

2 a Source: the surface of the wooden box rubbing on the bumps of the dirt road.
Reduced by: making the wooden box lighter.

b Source: the air molecules colliding with the parachute.
Reduced by: making the parachute smaller.

c Source: the surface of the end of the pump rubbing on the rubber of the valve.
Reduced by: putting some oil or other lubricant between the surfaces.

3 The cyclist has applied the brakes for an extended period of time. During the braking process, the kinetic energy of the bike is continuously converted to heat energy as a consequence of the friction between the disc and the brake pad.

4 The student could show her grandfather how fridge magnets move towards each other when released a small distance apart, how a towel falls to the floor when released and how a ruler rubbed on a jumper can be used to change the direction of a thin stream of water flowing from the tap when placed a short distance away.

5 The mass of an object is a measure of how much matter it comprises and is measured in kilograms. The weight of an object is a measure of the force due to gravity that acts on the object and is measured in newtons.

6

Conversion event	Converts energy from	Converts energy to
Headlights are turned on	Electrical energy	Light energy
Combustion engine is operating to drive car	Chemical energy	Kinetic energy
Car radio is turned on	Electrical energy	Sound energy
Boot opens automatically	Electrical energy	Gravitational potential energy/kinetic energy
Car heater is turned on	Chemical energy/electrical energy	Heat energy

7 When the combustion engine is operating, it converts chemical energy into heat energy and sound energy as well as the desired kinetic energy. Usually neither of these other forms is a desirable product for the engine and so some of the energy is effectively 'wasted'.

Chapter 4: Forces

WS 4.1 PAGE 57

1 You exert a contact force via your skin – you are touching the apple; that is, you are in contact with it.

2 The two components of the contact force are the normal force, which is perpendicular to the interface of the two surfaces, and the friction force, which is parallel to the interface.

3 The gravitational force – keeps Earth in orbit about the Sun.
The electrostatic force – makes your hair stand out and 'crackle' when brushed on a dry day.
The magnetic force – makes a magnet stick to a fridge door.

4 Newton's first law states that when no forces act on an object, the object will continue in its state of motion – so whatever velocity it is moving at (zero or otherwise) it will continue to move with that velocity. Note that Newton specifically referred to no forces, not to unbalanced forces.

5 Newton would say that there must be an external force acting on the ball to slow it down. We call that force *friction*.

6 $\vec{F}_{net} = m\vec{a}$ or $\vec{a} = \dfrac{\vec{F}_{net}}{m}$; the acceleration of an object, $\vec{a}$, is proportional to the total (net) force acting on the object, and inversely proportional to the mass of the object.

7 In a static situation, the acceleration of an object is zero, so $\vec{F}_{net} = m\vec{a} = 0$. The net force acting is zero. In a dynamic situation the net force is not zero, $\vec{F}_{net} = m\vec{a} \neq 0$, and so the object accelerates.

8 $\vec{F}_{a\,on\,b} = -\vec{F}_{b\,on\,a}$; when an object exerts a force on another object, it experiences an equal and opposite force – all forces are interactions.

9 The force that Emma applies to Narelle acts on Narelle, not on Emma. As long as the force Emma applies to Narelle is greater than any opposing forces, such as friction, Narelle will accelerate. Always remember that the two forces in Newton's third law act on different objects.

10 The force that Narelle applies to Emma and the force that Emma applies to Narelle are a Newton's third law pair: $\vec{F}_{N\,on\,E} = -\vec{F}_{E\,on\,N}$, and so they are always equal. It does not matter if one, or both, are accelerating.

11 The force that Narelle applies to Emma and the force that Emma applies to Narelle are a Newton's third law pair: $\vec{F}_{N\,on\,E} = -\vec{F}_{E\,on\,N}$, and so they are always equal. It doesn't matter if it is a static or dynamic situation.

12 a The gravitational force of the Moon on Earth:
$\vec{F}_{g,\,Moon\,on\,Earth} = -\vec{F}_{g,\,Earth\,on\,Moon}$

b The gravitational force of the book on Earth:
$\vec{F}_{g,\,book\,on\,Earth} = -\vec{F}_{g,\,Earth\,on\,book}$. Note that a normal force is *never* the Newton's third law pair to a gravitational force.

c The magnetic force of the fridge door on the magnet:
$\vec{F}_{magnet\,on\,fridge} = -\vec{F}_{fridge\,on\,magnet}$

d The friction force of the road on the tyre:
$\vec{F}_{road\,on\,tyre} = -\vec{F}_{tyre\,on\,road}$. This is the force that allows a car to accelerate forwards.

e The normal force of you on the floor:
$\vec{F}_{N,\,you\,on\,floor} = -\vec{F}_{N,\,floor\,on\,you}$

WS 4.2 PAGE 59

1 There are many examples. Any object moving with constant velocity, such as a skydiver falling at terminal velocity, is an example.

2 a Taking right to be positive:
$\vec{F}_{net} = \vec{F}_{Nick} + \vec{F}_{Ed} = -100\,N + 120\,N = 20\,N$ (to the right).

b The acceleration is in the direction of the net force: to the right.

c $\vec{F}_{net} = \vec{F}_{Nick} + \vec{F}_{Ed} = -120\,N + 120\,N = 0\,N$. Because the net force is zero, the acceleration is also zero.

d The net force, and hence acceleration, is zero, so the unicorn is now in equilibrium.

e The unicorn will not be stationary if it was moving before Nick increased his force, which is likely.

3 a The normal is greater than the gravitational force, to give a net upwards force.

b The normal is equal to the gravitational force, to give a zero net force.

c The normal is less than the gravitational force, to give a net downwards force.

4 a

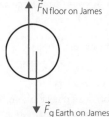

b

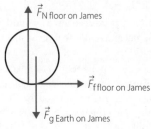

c

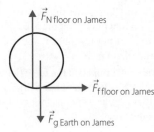

d

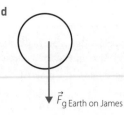

5 a

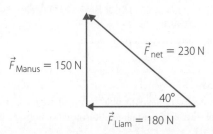

b $F_{net} = \sqrt{F_{Manus}^{\;2} + F_{Liam}^{\;2}} = \sqrt{(150\,N)^2 + (180\,N)^2} = 230\,N$

$\tan\theta = \dfrac{150\,N}{180\,N}, \theta = 40°$

c Your two answers should agree, within the uncertainty introduced by scale drawing. Using trigonometry is more precise, because although rounding errors may be introduced, these are generally far smaller than uncertainties introduced by drawing and measuring. Drawings should always be made to show the situation, but problems should be solved using trigonometry or algebra, not by scale drawing.

6 a

$\vec{F}_{net} = 26\,N$ $\vec{F}_{Anika} = 20\,N$

$51°$

$\vec{F}_{yabby} = 16\,N$

$F_{net} = \sqrt{F_{yabby}^{\;2} + F_{Anika}^{\;2}} = \sqrt{(16\,N)^2 + (20\,N)^2} = 26\,N$

$\tan\theta = \dfrac{20\,N}{16\,N}, \theta = 51°$

Note that we have arbitrarily put the force by the yabby to the left; it could just as well have been to the right.

Values are given to 2 significant figures to match the data in the question.

You could also solve this problem using scale drawing, but trigonometry is preferred because it is more precise.

b For the net force to be zero, a force of 26 N to the right and down, at 51° below the horizontal (an equal but opposite force to the combined forces from Anika and the first yabby).

c This is the Newton's third law pair to the force in part **b**, so it is 26 N up and left, at 51° above the horizontal.

7 $F_{net} = \sqrt{F_x^2 + F_y^2} = \sqrt{(270\,N)^2 + (450\,N)^2} = 520\,N$

$\tan\theta = \dfrac{450\,N}{270\,N}, \theta = 59°$

8 a $F_{net} = \sqrt{F_A^2 + F_B^2}$, so

$F_B = \sqrt{F_{net}^2 - F_A^2} = \sqrt{(250\,N)^2 - (200\,N)^2} = 150\,N$

You might have recognised this as a 3–4–5 triangle with hypotenuse $250\,N = 5 \times 50\,N$ and one side $4 \times 50\,N = 200\,N$, so the final side must be $3 \times 50\,N = 150\,N$. Using ratios like this is usually quicker and more precise (rounding errors are not introduced).

b, c $\vec{F}_{net} = 250\,N$ $\vec{F}_A = 200\,N$

$53°$

$\vec{F}_B = 150\,N$

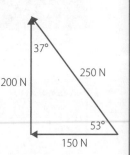

$\tan\theta = \dfrac{200\,N}{150\,N}, \theta = 53°$

For the other angle, remember that the three angles of a triangle add to 180°, so the last angle is $180° - 90° - 53° = 37°$.

Alternatively, you could use other trigonometric identities.

Note that you could also solve this problem using scale drawing, but this is less precise.

WS 4.3 PAGE 62

1 $\vec{F}_x = \vec{F}\cos\theta$

$\vec{F}_y = \vec{F}\sin\theta$

2 If the force is entirely in the vertical (y) direction, then the y component is the entire force: 300 N. In this case

$\vec{F}_x = \vec{F}\cos\theta = (300\,N)\cos 90° = 0$

$\vec{F}_y = \vec{F}\sin\theta = (300\,N)\sin 90° = 300\,N$

3 $\vec{F}_x = \vec{F}\cos\theta = (550\,N)\cos 35° = 451\,N$, so $\vec{F}_x = 450\,N$ (rounded)

$\vec{F}_y = \vec{F}\sin\theta = (550\,N)\sin 35° = 315\,N$, so $\vec{F}_y = 320\,N$ (rounded)

4 a For the shade sail to be in equilibrium, we require $\vec{F}_1 + \vec{F}_2 + \vec{F}_3 = 0$. Breaking into x and y components, and taking E as positive and N as positive:

$\vec{F}_{1,x} + \vec{F}_{2,x} + \vec{F}_{3,x} = 0$ so

$\vec{F}_{3,x} = -\vec{F}_{1,x} - \vec{F}_{2,x} = -100\,N - (-50\,N) = -50\,N$ or 50 N west

$\vec{F}_{1,y} + \vec{F}_{2,y} + \vec{F}_{3,y} = 0$ so

$\vec{F}_{3,y} = -\vec{F}_{1,y} - \vec{F}_{2,y} = -100\,N - (-100\,N) = 0\,N$, so no force in the north–south direction. $\vec{F}_3 = 50\,N$ west.

9780170449595

b

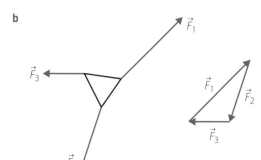

The forces should all add to zero, as seen in the vector addition diagram on the right above.

5   For each of these questions, the geometrical addition of the vectors is shown.

Note that you could solve these problems by drawing the vectors to scale and putting them tail to tip, and then measuring the resultant. However, this method is not recommended because unless your drawings are very precise, your answer will be very imprecise. The more vectors you have to add, the increasingly imprecise your answer will be. Hence trigonometry is always to be preferred to scale drawing.

a

$$\vec{F}_{net,x} = \vec{F}_{1,x} + \vec{F}_{2,x} = \vec{F}_1 \cos\theta_1 + \vec{F}_2 \cos\theta_2$$
$$= (3.0\,\text{N})\cos 90° + (5.0\,\text{N})\cos 40° = 3.8\,\text{N}$$
$$\vec{F}_{net,y} = \vec{F}_{1,y} + \vec{F}_{2,y} = \vec{F}_1 \sin\theta_1 + \vec{F}_2 \sin\theta_2$$
$$= (3.0\,\text{N})\sin 90° + (5.0\,\text{N})\sin 40° = 6.2\,\text{N}$$
$$F_{net} = \sqrt{F_{net,x}{}^2 + F_{net,y}{}^2} = \sqrt{(3.8\,\text{N})^2 + (6.2\,\text{N})^2} = 7.3\,\text{N}$$
$$\tan\theta = \frac{F_{net,y}}{F_{net,x}} = \frac{6.2\,\text{N}}{3.8\,\text{N}}, \theta = 59°$$

b

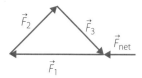

You need to pay attention to signs!

$$\vec{F}_{net,x} = \vec{F}_{1,x} + \vec{F}_{2,x} + \vec{F}_{3,x} = \vec{F}_1 \cos\theta_1 + \vec{F}_2 \cos\theta_2 + \vec{F}_3 \cos\theta_3$$
$$\vec{F}_{net,x} = (-5.0\,\text{N})\cos 0° + (3.0\,\text{N})\cos 45° + (3.0\,\text{N})\cos(-45°)$$
$$= -0.8\,\text{N}$$
$$\vec{F}_{net,y} = \vec{F}_{1,y} + \vec{F}_{2,y} + \vec{F}_{3,y} = \vec{F}_1 \sin\theta_1 + \vec{F}_2 \sin\theta_2 + \vec{F}_3 \sin\theta_3$$
$$\vec{F}_{net,y} = (-5.0\,\text{N})\sin 0° + (3.0\,\text{N})\sin 45° - (3.0\,\text{N})\sin 45° = 0\,\text{N}$$
$$= \sqrt{F_{net,x}{}^2 + F_{net,y}{}^2} = \sqrt{(-0.8\,\text{N})^2 + (0\,\text{N})^2} = 0.8\,\text{N}$$
$$\tan\theta = \frac{F_{net,y}}{F_{net,x}} = \frac{0\,\text{N}}{-0.8\,\text{N}}, \theta = 180° \text{ (the force is horizontal to the left)}$$

c

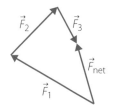

$$\vec{F}_{net,x} = \vec{F}_{1,x} + \vec{F}_{2,x} + \vec{F}_{3,x} = \vec{F}_1 \cos\theta_1 + \vec{F}_2 \cos\theta_2 + \vec{F}_3 \cos\theta_3$$
$$\vec{F}_{net,x} = (-4.0\,\text{N})\cos 30° + (3.0\,\text{N})\cos 45° + (1.5\,\text{N})\cos(60°)$$
$$= -0.6\,\text{N}$$
$$\vec{F}_{net,y} = \vec{F}_{1,y} + \vec{F}_{2,y} + \vec{F}_{3,y} = \vec{F}_1 \sin\theta_1 + \vec{F}_2 \sin\theta_2 + \vec{F}_3 \sin\theta_3$$
$$\vec{F}_{net,y} = (4.0\,\text{N})\sin 30° + (3.0\,\text{N})\sin 45° - (1.5\,\text{N})\sin 60° = 2.8\,\text{N}$$
$$F_{net} = \sqrt{F_{net,x}{}^2 + F_{net,y}{}^2} = \sqrt{(-0.6\,\text{N})^2 + (2.8\,\text{N})^2} = 2.9\,\text{N}$$
$$\tan\theta = \frac{F_{net,y}}{F_{net,x}} = \frac{2.8\,\text{N}}{-0.6\,\text{N}}, \theta = -78°$$

The negative sign indicates this is the angle from the negative side of the horizontal axis.

6   **a** $F_y = F\sin\theta$ so $\sin\theta = \dfrac{F_y}{F} = \dfrac{1650\,\text{N}}{3500\,\text{N}}$,

then $\theta = 28°$.

b   This can be done using Pythagoras's theorem or trigonometry.

$$F_x = \sqrt{F^2 - F_y^2} = \sqrt{(3500\,\text{N})^2 - (1650\,\text{N})^2} = 3086\,\text{N}$$
$$= 3100\,\text{N}$$

Note that we round to 2 significant figures as the least precise piece of data is given to two significant figures.

c $\vec{F}_g = m\vec{g} = (450\,\text{kg})(9.8\,\text{m s}^{-2}) = 4400\,\text{N}$ down (negative y)

d

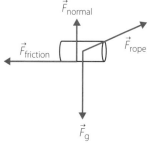

The contact force has two components, the friction force and the normal force.

e   The friction force acts in the horizontal direction and the normal force acts in the vertical. In both directions the sum of the forces is zero:

$$\vec{F}_{net,x} = \vec{F}_{rope,x} + \vec{F}_{friction,x} = 0 \text{ so}$$
$$\vec{F}_{friction,x} = -\vec{F}_{rope,x} = -3100\,\text{N (from part b)}$$
$$\vec{F}_{net,y} = \vec{F}_{rope,y} + \vec{F}_{normal} - \vec{F}_g = 0 \text{ so}$$
$$\vec{F}_{normal} = +\vec{F}_g - \vec{F}_{rope,y}$$
$$= 4400\,\text{N} - 1650\,\text{N} = 2750\,\text{N (2800 N, rounded)}$$

f   As a vector, $\vec{F}_{contact} = \vec{F}_{friction} + \vec{F}_{normal}$
which has magnitude

$$F_{contact} = \sqrt{F_{friction}{}^2 - F_{normal}{}^2}$$
$$= \sqrt{(-3100\,\text{N})^2 + (2750\,\text{N})^2}$$
$$= 4100\,\text{N}$$

1   There are various suitable inquiry questions, such as 'What is the maximum friction force the board can exert on …?', or 'Does the maximum friction force depend on the surface material of an object?'

3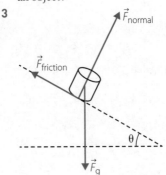

4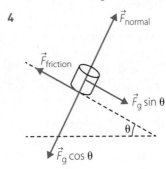

5   If we call the direction down the board the *x* direction, then before the object starts to slide,

$$\vec{F}_{net,\,x} = \vec{F}_g \sin\theta - \vec{F}_{friction} = 0 \text{ so } \vec{F}_{friction} = \vec{F}_g \sin\theta$$

6   You need to use the maximum value of angle θ before the object starts to slide, and the measured mass of the object. Take $g = 9.8 \, \text{m s}^{-2}$.

$$\vec{F}_{friction} = \vec{F}_g \sin\theta = mg \sin\theta$$

Do the calculation for each measured maximum angle.

8   a   Rough objects generally exert, and experience, a greater friction force, so need larger angles than smooth objects before they will slide.

    b   As we will see in the next chapter, the friction force depends on the normal force, which in general will be greater for heavier objects (when no other forces are acting). However, the angle will not depend on the mass, as both gravitational force and friction force scale linearly with mass in this experiment.

    c   What you can do to improve your experiment will depend on how you carried it out. In general you should always make repeat measurements to ensure reliability and control all variables other than the dependent variable to ensure validity.

Chapter 5: Forces, acceleration and energy

1   If the acceleration of an object is zero there no *net* force acting on it, so all the forces acting add to zero. It does *not* mean there are no forces at all.

2   There is not necessarily a force in the direction of motion. There may be no force at all, or only forces acting perpendicular to the direction of motion, such as an object sliding on a frictionless surface.

3   If an object is accelerating there must be a net force in the direction of the acceleration, but it does not necessarily have to be in the direction of motion. For example, when a car is braking

there is no force in the direction of motion, but only in the opposite direction.

4   For a car to move from rest it must accelerate, and the force to accelerate the car is the static friction force of the road on the tyres. This is also the force that provides braking and turning. The diagram below shows a car just accelerating from rest so there is not yet a significant air resistance.

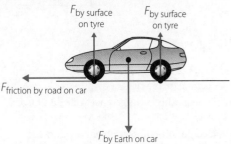

5   The static friction force is given by $F_{\text{static friction}} \le \mu_s F_N$.
So its minimum value is zero and this occurs when there is no external force trying to make one surface slide against the other. The maximum value is $F_{\text{static friction, max}} = \mu_s F_N$, which is $F_{\text{static friction, max}} = (0.75)(1500 \, \text{kg})(9.8 \, \text{m s}^{-2}) = 11 \, \text{kN}$ in this case.

This occurs just before the surfaces start to slide against each other, so the car is at maximum acceleration without skidding.

6   $$a_{\max} = \frac{F_{\max}}{m} = \frac{\mu_s F_N}{m} = \frac{\mu_s mg}{m}$$
$$= \mu_s g = (0.75)(9.8 \, \text{m s}^{-2}) = 7.4 \, \text{m s}^{-2}$$

7   The deceleration is limited by the friction between the tyres and road as well. So the same maximum acceleration applies. Converting from km h^{-1} to m s^{-1}:

$$100 \, \text{km h}^{-1} = 100 \ \text{km h}^{-1} \times \frac{1000 \, \text{m}}{1 \, \text{km}} \times \frac{1 \, \text{h}}{3600 \, \text{s}} = 27.8 \, \text{m s}^{-1}$$

Using the kinematics equation $v^2 = u^2 + 2as$ and recognising that $v = 0$:

$$0 = u^2 + 2as \text{ so } s = \frac{-u^2}{2a} = \frac{-(27.8 \, \text{m s}^{-1})^2}{2(-7.4 \, \text{m s}^{-2})} = 52 \, \text{m}$$

8   The coefficient of friction decreases, so the maximum deceleration also decreases. This means it takes a longer distance to come to a stop. As stopping distance increases with the square of the initial velocity, a small decrease in speed means a big difference in stopping distance.

9   When a car is skidding the tyres are sliding against the road. So the force the road exerts on the car is no longer static friction, but rather the force of kinetic friction. As this is lower, the maximum acceleration possible (whether braking or accelerating) is also lower, increasing the stopping distance.

10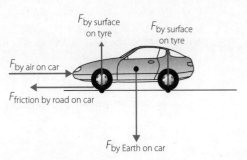

There are four forces acting on the car:
The gravitational force from Earth's gravitational field, acting on all parts of the car, but for simplicity drawn through the centre of mass.

The normal force of the road on the tyres, acting straight up, where the tyres touch the road. The total normal force is equal to the gravitational force.

The friction force of the road on the tyres, acting forwards (in the direction of motion), where the tyres touch the road.

The drag force due to the air, acting across all the front of the car. This force is the same magnitude as the friction force by the road.

11 The air resistance acting on the car means it would slow down if there were no friction force from the road pushing the car forwards. To create this friction force, the wheels need to turn to push against the road (Newton's third law). So the engine has to keep going to make the wheels keep pushing on the road.

12 The friction force on Neil is $F_{\text{kinetic friction}} = \mu_k F_N = \mu_k mg$.
Neil's acceleration is given by Newton's second law:
$$a = \frac{F}{m} = \frac{\mu_k mg}{m} = \mu_k g$$
Using the kinematics equation for constant acceleration,
$v^2 = u^2 + 2as$, so $v = \sqrt{u^2 + 2as}$
Substituting for a:
$$v = \sqrt{u^2 + 2\mu_k gs} = \sqrt{(3.5 \text{ m s}^{-1})^2 - 2(0.20)(9.8 \text{ m s}^{-2})(3.0 \text{ m})}$$
$$= 0.7 \text{ m s}^{-1}$$

13 We can go straight to our final equation and put in the new value for g:
$$v = \sqrt{u^2 + 2\mu_k g} = \sqrt{(3.5 \text{ m s}^{-1})^2 - 2(0.20)(1.6 \text{ m s}^{-2})(3.0 \text{ m})}$$
$$= 3.2 \text{ m s}^{-1}$$

WS 5.2 PAGE 71

1 When an object has a constant net force acting on it, its velocity increases linearly, so a graph of velocity versus time is a straight line with gradient given by $\frac{F_{\text{net}}}{m}$.

2 a Constant speed in a straight line means constant velocity, which means zero acceleration and hence zero net force – so the net force is constant and zero.

b The direction of motion is changing, so there is an acceleration, and the direction of the acceleration is also changing – so there is a non-zero net force, which is *not* constant. Remember that just because speed is constant, it doesn't mean zero force! The magnitude of a force may be constant even if its direction is not.

c If the acceleration is constant, as stated, then by Newton's second law the net force is also constant.

d In this case the acceleration is decreasing, so the net force must also be decreasing, hence it is not constant.

3 a
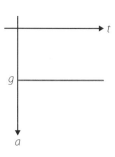

b
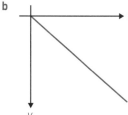

c
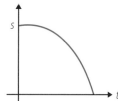

4 a $F_{\text{static friction, max}} = \mu_s N = \mu_s mg = -(0.80)(1200 \text{ kg})(9.8 \text{ m s}^{-2})$
$= -9408 \text{ N} = -9400 \text{ N}$ to two significant figures.
The negative sign comes from defining the positive direction as the direction of motion. This force acts in the opposite direction.

b $a_{\text{max}} = \frac{F_{\text{max}}}{m} = \frac{-9400 \text{ N}}{1200 \text{ kg}} = -7.8 \text{ m s}^{-2}$
The negative sign is because the force, and hence acceleration, is in the negative direction.

c We can use $a = \frac{v - u}{t}$, and note that the final speed v is zero, so
$$t = \frac{v - u}{a} = \frac{0 - 10 \text{ m s}^{-1}}{-7.8 \text{ m s}^{-2}} = 1.3 \text{ s}$$

d Use the kinematics equation:
$$s = ut + \frac{1}{2} at^2 = (10 \text{ m s}^{-1})(1.3 \text{ s}) - \frac{1}{2} (7.8 \text{ m s}^{-2})(1.3 \text{ s})^2$$
$$= 6.4 \text{ m}$$

e Now we use $t = 0.65 \text{ s}$:
$$s = ut + \frac{1}{2} at^2 = (10 \text{ m s}^{-1})(0.65 \text{ s}) - \frac{1}{2} (7.8 \text{ m s}^{-2})(0.65 \text{ s})^2$$
$$= 4.8 \text{ m}$$
Note that this is far more than half the distance.

f i
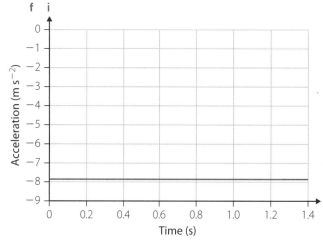

ii
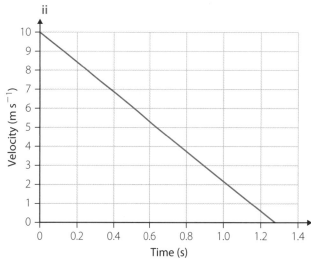

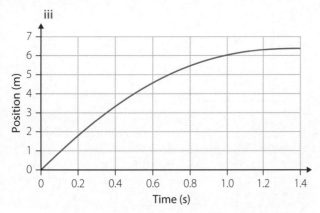

5 Air resistance is not a constant force, but varies with the speed of the object. So we cannot use kinematics equations that assume constant acceleration, because that implies constant force, to model the motion of objects that have air resistance acting on them.

6 a The direction and magnitude are given by Newton's second law:

$$a = \frac{F_{net}}{m} = \frac{250\,\text{N}}{20\,\text{kg}} = 12.5\,\text{m s}^{-2}$$

The direction is the same as that of the force, at 45° above the horizontal.

b i   **ii**

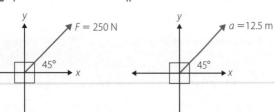

7 a

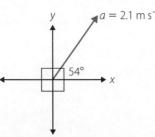

The net force in the x direction is:

$$\vec{F}_{net,x} = \vec{F}_{1,x} + \vec{F}_{2,x} = \vec{F}_1 \cos\theta_1 + \vec{F}_2 \cos\theta_2$$
$$= (1.0\,\text{N})\cos 90° + (2.5\,\text{N})\sin 40° = 2.6\,\text{N}$$

The net force in the y direction is:

$$\vec{F}_{net,y} = \vec{F}_{1,y} + \vec{F}_{2,y} = \vec{F}_1 \sin\theta_1 + \vec{F}_2 \sin\theta_2$$
$$= (1.0\,\text{N})\sin 90° + (2.5\,\text{N})\cos 40° = 1.9\,\text{N}$$

So the total force is

$$F_{net} = \sqrt{F_{net,x}^2 + F_{net,y}^2} = \sqrt{(1.9\,\text{N})^2 + (2.6\,\text{N})^2} = 3.2\,\text{N}$$

$$\tan\theta = \frac{F_y}{F_x} = \frac{2.6\,\text{N}}{1.9\,\text{N}}, \theta = 54° \text{ to the } x\text{-axis.}$$

The acceleration will be at this same angle, and will have magnitude

$$a = \frac{F_{net}}{m} = \frac{3.2\,\text{N}}{1.5\,\text{kg}} = 2.1\,\text{m s}^{-2}$$

Note that we could also have found the x and y components of the acceleration from the force components and then added them.

b

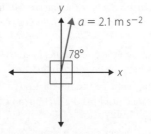

The net force in the x direction is:

$$\vec{F}_{net,x} = \vec{F}_{1,x} + \vec{F}_{2,x} + \vec{F}_{3,x} = \vec{F}_1 \cos\theta_1 + \vec{F}_2 \cos\theta_2 + \vec{F}_3 \cos\theta_3$$
$$= (-3.0\,\text{N})\cos 45° - (1.5\,\text{N})\cos 60° + (4.0\,\text{N})\cos 30° = 0.6\,\text{N}$$

The net force in the y direction is:

$$\vec{F}_{net,y} = \vec{F}_{1,y} + \vec{F}_{2,y} + \vec{F}_{3,y} = \vec{F}_1 \sin\theta_1 + \vec{F}_2 \sin\theta_2 + \vec{F}_3 \sin\theta_3$$
$$= (3.0\,\text{N})\sin 45° - (1.5\,\text{N})\sin 60° + (4.0\,\text{N})\sin 30° = 2.8\,\text{N}$$

So the total force is:

$$F_{net} = \sqrt{F_{net,x}^2 + F_{net,y}^2} = \sqrt{(0.6\,\text{N})^2 + (2.8\,\text{N})^2} = 2.9\,\text{N}$$

$$\tan\theta = \frac{F_y}{F_x} = \frac{2.8\,\text{N}}{0.6\,\text{N}}, \theta = 78° \text{ to the } x\text{-axis.}$$

The acceleration will be at this same angle, and will have magnitude

$$a = \frac{F_{net}}{m} = \frac{2.9\,\text{N}}{1.5\,\text{kg}} = 1.9\,\text{m s}^{-2}$$

8 The first thing to do is draw a force diagram for Marcus and the blanket. Note that we treat them as a single object.

First we draw a force diagram showing all the individual forces acting:

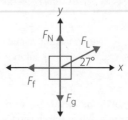

Then we break the forces into components in the x and y directions:

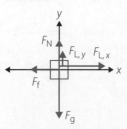

It is very important that you *do not* assume that the normal force, F_N, is equal to the gravitational force, F_g. The normal force is reduced because of the force exerted by Laurence, F_L.

The net force, and hence the acceleration, is in the x direction, because Marcus is sliding along the floor, not being lifted off it. The net force in the y direction is therefore zero. We can use this to find the normal force, F_N:

$$F_{L,y} + F_N - F_g = 0$$

So

$$F_N = F_g - F_{L,y} = m_M g - F_L \sin 27°$$
$$= (35\,\text{kg})(9.8\,\text{N kg}^{-1}) - (120\,\text{N})(0.454) = 289\,\text{N}$$

Now we can find the net force acting in the x direction:

$$F_{net} = F_{L,x} - \mu_k F_N = F_L \cos 27° - (0.15)(289\,\text{N}) = 63.6\,\text{N}$$

And finally, we can use Newton's second law to find Marcus's acceleration:

$$a = \frac{F}{m} = \frac{63.6\,\text{N}}{35\,\text{kg}} = 1.8\,\text{m s}^{-2}$$

WS 5.3 PAGE 76

1 Kinetic energy is energy associated with the motion of an object or its parts. This can be translation, or rotation, or the motion of a wave through a material.

2 $E_k = \frac{1}{2}mv^2$ where E_k is the kinetic energy of an object with mass m moving at speed v.

3 Kinetic energy cannot be negative because mass cannot be negative, and while velocity can be negative, the square of velocity must always be positive.

4 a Convert to SI units:

$$100\,\text{km h}^{-1} = 100\,\text{km h}^{-1} \times \frac{1000\,\text{m}}{1\,\text{km}} \times \frac{1\,\text{h}}{3600\,\text{s}} = 27.8\,\text{m s}^{-1}$$

Then: $E_k = \frac{1}{2}mv^2 = \frac{1}{2}(100\,\text{kg})(27.8\,\text{m s}^{-1})^2 = 580\,\text{kJ}$.

b $E_k = \frac{1}{2}mv^2$ so $v = \sqrt{\frac{2E_k}{m}} = \sqrt{\frac{2(210\,000\,\text{J})}{1500\,\text{kg}}} = 16.7\,\text{m s}^{-1}$

Convert to km h^{-1}:

$$16.7\,\text{m s}^{-1} = 16.7\,\text{m s}^{-1} \times \frac{1\,\text{km}}{1000\,\text{m}} \times \frac{3600\,\text{s}}{1\,\text{h}} = 60\,\text{km h}^{-1}.$$

5 Potential energy is energy stored in a system due to the forces acting between the objects in the system. It is energy that has the potential to be converted into kinetic energy.

6 $U_g = mgh$ where U_g is the gravitational potential energy, m is the mass of the object, g is the acceleration due to gravity close to Earth's surface and h is the height above a convenient refence point – usually (but not always) the ground.

7 If the reference height is greater than the height h in the equation for U_g then U_g will be negative. For example, if we take the starting height of a falling object as zero, then its initial gravitational potential energy is zero and becomes increasingly negative as it falls.

8 a $U_g = mgh = (1.5\,\text{kg})(9.8\,\text{m s}^{-2})(3.5\,\text{m}) = 51\,\text{kg m}^2\,\text{s}^{-2}$
$= 51\,\text{J}$

b We are taking the zero of gravitational potential energy as 2.4 m above the ground.
$U_g = mgh = (1.5\,\text{kg})(9.8\,\text{m s}^{-2})(-2.4\,\text{m}) = -35\,\text{kg m}^2\,\text{s}^{-2} = -35\,\text{J}$

9 Mechanical energy refers to kinetic energy of whole objects (not their component atoms, which is thermal energy) and potential energies including gravitational potential energy and energy stored in springs. Mechanical energy does not include chemical potential energy or thermal energy.

10 Mechanical energy is conserved in an isolated system (one which has no movement of matter or energy in or out) when there are no frictional forces, including air resistance, acting.

11 Work, W, is the energy transferred to an object due to the action of a force on the object, as the object moves through some displacement. Work has the same units as energy: J.
$W = Fs\cos\theta$, where F is the force, s is the displacement and θ is the angle between the force and the displacement.

12

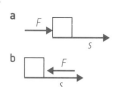

13

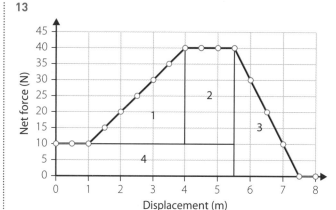

The work is the area under the line. We can count squares and multiply by the area of each square, or divide the shape into rectangles and triangles as shown.

$$\text{Area } 1 = \frac{1}{2} \times \text{base} \times \text{height}$$
$$= \frac{1}{2} \times 3\,\text{m} \times 30\,\text{N} = 45\,\text{N m} = 45\,\text{J}$$

$$\text{Area } 2 = \text{base} \times \text{height} = 1.5\,\text{m} \times 30\,\text{N} = 45\,\text{N m} = 45\,\text{J}$$

$$\text{Area } 3 = \frac{1}{2} \times \text{base} \times \text{height}$$
$$= \frac{1}{2} \times 2\,\text{m} \times 40\,\text{N} = 40\,\text{N m} = 40\,\text{J}$$

$$\text{Area } 4 = \text{base} \times \text{height} = 5.5\,\text{m} \times 10\,\text{N} = 55\,\text{N m} = 55\,\text{J}$$

Total work $= 45\,\text{J} + 45\,\text{J} + 40\,\text{J} + 55\,\text{J} = 185\,\text{J}$

14 a $1\,\text{km} = 1000\,\text{m}$
$W = Fs\cos\theta = (500\,\text{N})(1000\,\text{m})\cos 25° = 453\,\text{kJ} = 450\,\text{kJ}$.

b The work–kinetic energy theorem says that the work done on an object is converted to kinetic energy of the object. As the car is starting from rest:

$$E_k = \frac{1}{2}mv^2 = W$$

Rearranging for v:

$$v = \sqrt{\frac{2W}{m}} = \sqrt{\frac{2(450\,000\,\text{J})}{1800\,\text{kg}}} = 22\,\text{m s}^{-1}$$

Note that $22\,\text{m s}^{-1}$ is approximately $80\,\text{km h}^{-1}$ so this is unrealistically fast – there will also be rolling friction and air resistance in the horizontal direction.

15 Friction acts in the direction opposite the displacement, so it does negative work. We can also conclude this from the fact that the book is slowing down.

16 Friction can do positive work. The static friction force on car tyres does positive work when the car is accelerating or at constant speed. When you walk, the friction force of the ground on your feet does positive work on you. Whenever the friction force is in the direction of displacement it does positive work.

17 Gravitational potential energy is converted to kinetic energy. The initial gravitational potential energy is $U_{g,\,\text{initial}} = mgh_{\text{initial}}$, and the final gravitational potential energy is zero. The kinetic energy starts at zero and increases to $E_{k,\text{final}} = \frac{1}{2}mv^2$.

So $mgh_{\text{initial}} = \frac{1}{2}mv_{\text{final}}^2$, so

$$v_{\text{final}} = \sqrt{2gh} = \sqrt{2(9.8\,\text{m s}^{-2})(3.5\,\text{m})} = 8.3\,\text{m s}^{-1}.$$

18 a We can use conservation of mechanical energy to solve this. The total vertical distance is
$y = (12\,\text{m})\sin 35° = 6.9\,\text{m}$.

$mgh_{\text{initial}} = \dfrac{1}{2}mv_{\text{final}}^2$ so

$v_{\text{final}} = \sqrt{2gh} = \sqrt{2(9.8\,\text{ms}^{-2})(6.9\,\text{m})} = 11\,\text{ms}^{-1}$.

b Ignoring friction, so we can make the approximation of conservation of mechanical energy, the speed will be the same because the start and end points are the same.

c The friction force acts for less distance on the straight slide, so does less negative work on the book. Hence less mechanical energy is lost (converted to thermal energy) in this case so the book hits the ground faster from the straight slide.

WS 5.4 PAGE 80

1 Power is the rate at which work is done, or energy is transferred. It has units of $\text{J s}^{-1} = \text{W}$.

2 $1.5\,\text{kW} = 1500\,\text{W}$, 1 minute = 60 s

$P = \dfrac{\Delta E}{t}$ so $\Delta E = Pt = 1500\,\text{W} \times 60\,\text{s} = 90\,\text{kJ}$

3 Convert to SI units:

$100\,\text{km h}^{-1} = 100\,\text{km h}^{-1} \times \dfrac{1000\,\text{m}}{1\,\text{km}} \times \dfrac{1\,\text{h}}{3600\,\text{s}} = 27.8\,\text{m s}^{-1}$

$P = \dfrac{\Delta E}{t} = \dfrac{\Delta E_k}{t}$. The initial E_k is zero, and the final E_k is

$E_k = \dfrac{1}{2}mv_{\text{final}}^2$

so $P = \dfrac{\Delta E}{t} = \dfrac{mv_{\text{final}}^2}{2t} = \dfrac{(1200\,\text{kg})(27.8\,\text{m s}^{-1})^2}{2(11\,\text{s})} = 42\,\text{kW}$

4 We want the time for Neil's kinetic energy to drop to zero.

$P = \dfrac{\Delta E}{t} = \dfrac{\Delta E_k}{t}$, so $t = \dfrac{\Delta E_k}{P}$. The final E_k is zero, and the initial

E_k is $E_k = \dfrac{1}{2}mv_{\text{initial}}^2$

$t = \dfrac{\Delta E_k}{P} = \dfrac{mv_{\text{initial}}^2}{2P} = \dfrac{(48\,\text{kg})(2.0\,\text{m s}^{-1})^2}{2(38\,\text{W})} = 2.5\,\text{s}$

5 $P = \dfrac{\Delta E}{t} = Fv$. The gravitational force is constant, so the book has constant acceleration, which means the velocity of the book increases linearly.

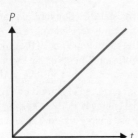

6 a If the lift is moving at constant velocity then there is zero net force on the lift. So the force by the motor must be equal in magnitude (but opposite in direction) to the gravitational force.
$F_{\text{lift}} = F_g = mg = (750\,\text{kg})(9.8\,\text{m s}^{-2}) = 7350\,\text{N}$ or $7400\,\text{N}$ rounded correctly.

b $\Delta U_g = mg\Delta h = (7350\,\text{N})(10\,\text{m}) = 73\,500\,\text{J}$

c $W = Fs\cos\theta = (7350\,\text{N})(10\,\text{m})\cos 0° = 73\,500\,\text{J}$
Note that this is the same (as it must be) as the work done by the gravitational force because the speed is constant.

d $P = \dfrac{\Delta E}{t} = \dfrac{73\,500\,\text{J}}{12\,\text{s}} = 6125\,\text{W}$ or $6.1\,\text{kW}$

e The force is the same, but the power is greater.

$P = Fv$ so $v = \dfrac{P}{F} = \dfrac{10\,000\,\text{W}}{7350\,\text{N}} = 1.4\,\text{m s}^{-1}$.

7 a We can use the idea of work. The work done is the change in kinetic energy,

$\Delta E_k = E_{k,\text{final}} - E_{k,\text{initial}}$ where $E_{k,\text{final}} = 0$ and

$E_{k,\text{initial}} = \dfrac{1}{2}mv_{\text{initial}}^2$

Work is also $W = Fs\cos\theta = -Fs$ since $\cos\theta = 1$ and the force is in the opposite direction to the displacement.

Putting it together: $W = -Fs = -\dfrac{1}{2}mv_{\text{initial}}^2$, so

$F = \dfrac{mv_{\text{initial}}^2}{2s} = \dfrac{(0.25\,\text{kg})(1.5\,\text{m s}^{-1})^2}{2(5.0\,\text{m})} = 0.056\,\text{N}$

b We can use kinematics and Newton's second law to work out how long it takes the ball to stop, then use $P = \dfrac{\Delta E}{t}$ where $\Delta E = E_{k,\text{final}} - E_{k,\text{initial}}$. But there is a quicker way: we can use the average force calculated above, and the average velocity. If the force is constant, then the average velocity is simply $\dfrac{1}{2}v_{\text{initial}} = 0.75\,\text{m s}^{-1}$.

Then $P_{\text{ave}} = Fv_{\text{ave}} = (0.056\,\text{N})(0.75\,\text{m s}^{-1}) = 0.042\,\text{W}$ or $42\,\text{mW}$. Try doing this question the other way to confirm you get the same answer!

8 a The velocity is constant, so the force of friction on the car's tyres (pushing it forwards) is equal to the air resistance (pushing it back). $F_{\text{friction}} = 350\,\text{N}$.

b Convert to SI units:

$80\,\text{km h}^{-1} = 80\,\text{km h}^{-1} \times \dfrac{1000\,\text{m}}{1\,\text{km}} \times \dfrac{1\,\text{h}}{3600\,\text{s}} = 22.2\,\text{m s}^{-1}$

$P = Fv = (350\,\text{N})(22.2\,\text{m s}^{-1}) = 7.8\,\text{kW}$

c $P = Fv$ so $v = \dfrac{P}{F} = \dfrac{2100\,\text{W}}{150\,\text{N}} = 14\,\text{m s}^{-1}$

Convert to km h^{-1}:

$14\,\text{m s}^{-1} = 14\,\text{m s}^{-1} \times \dfrac{1\,\text{km}}{1000\,\text{m}} \times \dfrac{3600\,\text{s}}{1\,\text{h}} = 50\,\text{km h}^{-1}$

9 a This is the horizontal component of the force exerted by Laurence, as Marcus is sliding along the floor, not going up or down.

$F_x = F_L\sin\theta = (150\,\text{N})\sin 27° = 68\,\text{N}$

b We can use the force to find the acceleration, and then use kinematics to find speed. From the speed we can then find the kinetic energy.

$a = \dfrac{F_{\text{net}}}{m} = \dfrac{F_x}{m}$

$v = u + at = 0 + at = \dfrac{F_x t}{m}$

$E_k = \dfrac{1}{2}mv^2 = \dfrac{1}{2}m\left(\dfrac{F_x t}{m}\right)^2 = \dfrac{1}{2}\dfrac{F_x^2 t^2}{m}$

9780170449595

Before we substitute numbers we should check that our expression is dimensionally correct:

$$\frac{F_x^2 t^2}{m} = \frac{N^2 s^2}{kg} = \frac{(kg\,m\,s^{-2})^2 s^2}{kg} = kg\,m^2\,s^{-2} = J$$

This is the correct unit for energy, so we can now substitute in the values to find the answer. Note that this preferable to calculating numbers along the way because it gives us some idea of how the variables are related – we can see, for example, that the change in kinetic energy is proportional to the square of the applied force over time.

Substituting values:

$$E_k = \frac{1}{2}\frac{F_x^2 t^2}{m} = \frac{1}{2}\frac{(68\,N)^2 (1s)^2}{35\,kg} = 66\,J$$

c $P = \dfrac{\Delta E}{t} = \dfrac{66\,J}{1s} = 66\,W$

Chapter 6: Momentum, energy and simple systems

WS 6.1 **PAGE 83**

1   Possible inquiry questions include 'Is momentum conserved in one-dimensional collisions?', 'How does the velocity of a moving marble change when it collides with a stationary one?', and 'Does the change depend on the relative sizes of the two marbles?' There are many other possible questions.

2

What are the risks in doing this experiment?	How can you manage these risks to stay safe?
Standing on a marble could lead to a fall	Take care not to drop any marbles on the floor.

3   If the two marbles are identical then we expect them to 'swap' velocities, and the second marble to move off with the initial speed of the first marble. The first marble is then left behind stationary. However it is more likely that, due to friction, including friction between the marbles which are rotating, either there will be slight recoil or incomplete momentum transfer.

4,7,9   The table shows how results should be calculated for each of questions 4, 7, 9.

	Marble 1, marble 2	System (marble 1 + marble 2)
Mass (kg)	Measured with scales	
Time 1 (s)	Measured with stop watch	
Position 1 (m)	Measured with ruler	
Time 2 (s)		
Position 2 (m) (position at time 2)		
Velocity before collision (m s⁻¹)	Calculated from equation for v	
Momentum before collision (kg m s⁻¹)	Calculated from equation for p	Sum of the two momenta
Kinetic energy before the collision (J)	Calculated from equation for E_k	Sum of the two kinetic energies

~continued in right column ▲

	Marble 1, marble 2	System (marble 1 + marble 2)
Time 3 (s)		
Position 3 (m)		
Time 4 (s)		
Position 4 (m)		
Velocity after collision (m s⁻¹)		
Momentum after collision (kg m s⁻¹)		Sum of the two momenta, noting that they may have opposite signs
Kinetic energy after the collision (J)		Sum of the two kinetic energies

5   $v = \dfrac{s_2 - s_1}{t_2 - t_1}$, $p = mv$, $E_k = \dfrac{1}{2}mv^2$

6   We expect the large marble to transfer some, but not all, of its momentum and kinetic energy to the small marble, and both to move off in the direction of initial motion.

8   We expect the small marble to transfer some of its momentum and kinetic energy to the large marble, and to bounce back (recoil) as the large marble moves off in the direction of initial motion.

10 a   The momentum of the system is constant for an isolated system. However, there is friction from the surface of the table and the rulers, so the system is not perfectly isolated. Hence we expect to see a slight decrease in momentum.

     Similarly, the kinetic energy of the system of marbles will decrease because not only is there friction due to the table and rulers, but also the collisions themselves are not perfectly elastic – kinetic energy is lost to sound and thermal energy during each collision.

   b   Using a short time period minimises the effects of the frictional forces from the table and rulers.

   c   How well the marbles approximated an isolated system depends on how close to zero the change in total momentum was: the closer to zero Δp is, the better the approximation to an isolated system.

   d   The experiment could be made more reliable by strategies such as making repeat measurements and using more precise equipment.

   e   An experiment is valid when all variables (other than the dependent variable) are controlled.

   f   There are many ways to extend this experiment, for example allowing both marbles to be moving before the collision, or using more than two marbles.

WS 6.2 **PAGE 88**

1   $10\,g = 0.01\,kg$, $20\,cm\,s^{-1} = 0.2\,m\,s^{-1}$
    $p = mv = (0.01\,kg)(0.2\,m\,s^{-1}) = 0.002\,kg\,m\,s^{-1}$

2   $p = mv$ so $v = \dfrac{p}{m} = \dfrac{23\,000\,kg\,m\,s^{-1}}{1500\,kg} = 15.3\,m\,s^{-1}$

    Convert to $km\,h^{-1}$ units:

    $15.3\,m\,s^{-1} = 15.3\,m\,s^{-1} \times \dfrac{1\,km}{1000\,m} \times \dfrac{3600\ s}{1h} = 55\,km\,h^{-1}$

3 $p = mv$ and $E_k = \dfrac{1}{2}mv^2$

Multiply E_k by $\dfrac{m}{m}$: $E_k = \dfrac{1}{2}\dfrac{m^2v^2}{m}$

Substitute $mv = p$: $E_k = \dfrac{p^2}{2m}$

4 $58\,g = 0.058\,kg$

The quick way to do this problem is to use the expression derived above.

$E_k = \dfrac{p^2}{2m} = \dfrac{(0.46\,kg\,m\,s^{-1})^2}{2(0.058\,kg)} = 1.8\,kg\,m^2\,s^{-2} = 1.8\,J$

Alternatively, you can use $p = mv$ to calculate the velocity, then use $E_k = \dfrac{1}{2}mv^2$ to calculate the kinetic energy. Do the problem both ways and check that you get the same answer.

5 Again, the quick way to do this problem is to use the expression derived above.

$E_k = \dfrac{p^2}{2m}$ so

$p = \sqrt{2mE_k} = \sqrt{2(0.058\,kg)(0.75\,J)} = 0.29(kg\,J)^{\frac{1}{2}}$

$= 0.29(kg^2\,m^2\,s^{-2})^{\frac{1}{2}} = 0.29\,kg\,m\,s^{-1}$

Alternatively, you can use $E_k = \dfrac{1}{2}mv^2$ to calculate the velocity, then use to $p = mv$ calculate the momentum. Try doing this to check your answer.

6 a $E_k = \dfrac{1}{2}mv^2$ so $v = \sqrt{\dfrac{2E_k}{m}}$

E_k is the same for both animals, but if $m_{cat} = m$, then $m_{dog} = 2m$.

$v_{cat} = \sqrt{\dfrac{2E_k}{m}}$, $v_{dog} = \sqrt{\dfrac{2E_k}{2m}}$

$\dfrac{v_{cat}}{v_{dog}} = \sqrt{2}$ or $v_{cat} = \sqrt{2}v_{dog}$

b Momentum is given by $p = mv$.

For the cat: $p_{cat} = m\sqrt{2}v_{dog} = \sqrt{2}mv_{dog}$

For the dog: $p_{dog} = 2mv_{dog}$

$2 > \sqrt{2}$ so the dog has the greater momentum.

7 a In a perfectly elastic collision, kinetic energy is conserved, so all the kinetic energy the ball has just before the collision, it still has after the collision, if we ignore any energy change to Earth. Ignoring air resistance, all the kinetic energy of the ball is converted to gravitational potential energy as it goes up, so it reaches the same height as it fell from: 1.3 m.

b We could use kinematics, the idea of work, or conservation of energy.

Using conservation of energy

$E_{k, final} = \dfrac{1}{2}mv^2 = mgh = U_{g, initial}$, so

$v = \sqrt{2gh} = 5.0\,m\,s^{-1}$

c Taking down as the negative direction:

$\Delta p = p_{final} - p_{initial} = mv_{final} - mv_{initial}$

$= (0.1\,kg)(5.0\,m\,s^{-1}) - (0.1\,kg)(-5.0\,m\,s^{-1})$

$= 1\,kg\,m\,s^{-1}$ upwards, as this is positive.

d $E_k = \dfrac{1}{2}mv^2 = \dfrac{1}{2}(0.1\,kg)(5.0\,m\,s^{-1})^2 = 1.25\,kg\,m^2\,s^{-2} = 1.3\,J$

e The kinetic energy does not change, as we are assuming a perfectly elastic collision and ignoring the change in kinetic energy of Earth.

f Momentum is conserved, so if the ball's momentum has changed by $1\,kg\,m\,s^{-1}$ upwards, then Earth has changed momentum by this much in the opposite direction (downwards).

g $p = mv$ so $v = \dfrac{p}{m} = \dfrac{1\,kg\,m\,s^{-1}}{6.0 \times 10^{24}\,kg} = 1.7 \times 10^{-25}\,m\,s^{-1}$

h $E_k = \dfrac{1}{2}mv^2 = \dfrac{1}{2}(6.0 \times 10^{24}\,kg)(1.7 \times 10^{-25}\,m\,s^{-1})^2$

$= 8.7 \times 10^{-26}\,J$

i The change in the kinetic energy of Earth is 27 orders of magnitude smaller than the kinetic energy if the ball before the collision, so the fraction of the ball's kinetic energy that is lost to Earth is negligible and we can justify treating Earth as stationary both before and after the collision.

8 a If we treat the wall as part of Earth and assume that its change in kinetic energy is negligible, then kinetic energy is conserved in this collision.

b Momentum is always conserved in any interaction. It does not appear to be conserved in this case if we consider only the marble. But the wall/Earth also has a momentum change, and this must be equal and opposite to the momentum change of the marble.

c As both momentum and kinetic energy are conserved, this is an elastic collision.

9 We can calculate the velocity using conservation of momentum.

$p_i = p_f$, so $m_1u_1 + m_2u_2 = m_1v_1 + m_2v_2$.

Since the marbles have the same mass, this reduces to:

$u_1 + u_2 = v_1 + v_2$, so

$v_2 = u_1 + u_2 - v_1 = 5.5\,m\,s^{-1} + 2.5\,m\,s^{-1} - 2.5\,m\,s^{-1}$

$v_2 = 5.5\,m\,s^{-1}$

10 We can calculate the velocity using conservation of momentum:

$p_i = p_f$, so $m_1u_1 + m_2u_2 = m_1v_1 + m_2v_2$

Given $m_1 = 2m_2$, this reduces to:

$2u_1 + u_2 = 2v_1 + v_2$

$v_2 = 2u_1 + u_2 - 2v_1 = 2(5.5\,m\,s^{-1}) + 2.5\,m\,s^{-1} - 2(3.5\,m\,s^{-1})$

$v_2 = 6.5\,m\,s^{-1}$

11 a Momentum and kinetic energy are conserved in elastic collisions. We could write expressions for momentum and kinetic energy before and after, and then solve them as simultaneous equations. But this is a bit messy, so instead we use the derived equations (page 163 of the Student Book):

(1) $u_1 + v_1 = u_2 + v_2$ and (2) $m_1u_1 + m_2u_2 = m_1v_1 + m_2v_2$

From (1): $2.0\,m\,s^{-1} + v_1 = 0\,m\,s^{-1} + v_2$ so $v_2 = 2.0\,m\,s^{-1} + v_1$

From (2): $m(2.0\,m\,s^{-1}) + 0 = mv_1 + 1.5mv_2$

so $v_1 = 2.0\,m\,s^{-1} - 1.5v_2$

Substitute the expression for v_2 from the equation derived from (1):

$v_1 = 2.0\,m\,s^{-1} - 1.5v_2 = 2.0\,m\,s^{-1} - 1.5(2.0\,m\,s^{-1} + v_1)$

$= -1.0\,m\,s^{-1} - 1.5v_1$

so $2.5v_1 = -1.0\,m\,s^{-1}$ and $v_1 = -0.4\,m\,s^{-1}$. The negative sign means it has bounced back.

$v_2 = 2.0\,m\,s^{-1} + v_1 = 2.0\,m\,s^{-1} + (-0.4\,m\,s^{-1}) = 1.6\,m\,s^{-1}$

b (1) $u_1 + v_1 = u_2 + v_2$ and (2) $m_1u_1 + m_2u_2 = m_1v_1 + m_2v_2$

From (1): $2.0\,m\,s^{-1} + v_1 = 1.5\,m\,s^{-1} + v_2$ so

$v_2 = 0.5\,m\,s^{-1} + v_1$

From (2): $m(2.0\,m\,s^{-1}) + 1.5m(1.5\,m\,s^{-1}) = mv_1 + 1.5mv_2$

so $4.25\,m\,s^{-1} = v_1 + 1.5v_2$ and $v_1 = 4.25\,m\,s^{-1} - 1.5v_2$

Substitute the expression for v_2 from the equation derived from (1):

$v_1 = 4.25\,\text{m s}^{-1} - 1.5v_2 = 4.25\,\text{m s}^{-1} - 1.5(0.5\,\text{m s}^{-1} + v_1)$
$\quad = 3.5\,\text{m s}^{-1} - 1.5v_1$

so $2.5v_1 = 3.5\,\text{m s}^{-1}$ and $v_1 = 1.4\,\text{m s}^{-1}$ (to the right).
$v_2 = 0.5\,\text{m s}^{-1} + v_1 = 1.9\,\text{m s}^{-1}$

c (1) $u_1 + v_1 = u_2 + v_2$ and (2) $m_1u_1 + m_2u_2 = m_1v_1 + m_2v_2$

From (1): $2.0\,\text{m s}^{-1} + v_1 = -1.5\,\text{m s}^{-1} + v_2$ so $v_2 = 3.5\,\text{m s}^{-1} + v_1$
From (2): $m(2.0\,\text{m s}^{-1}) + 1.5m(-1.5\,\text{m s}^{-1}) = mv_1 + 1.5mv_2$
so $-0.5\,\text{m s}^{-1} = v_1 + 1.5v_2$ and $v_1 = -0.5\,\text{m s}^{-1} - 1.5v_2$

Substitute the expression for v_2 from the equation derived from (1):

$v_1 = -0.5\,\text{m s}^{-1} - 1.5v_2 = 4v_1 = -0.5\,\text{m s}^{-1} - 1.5(3.5\,\text{m s}^{-1} + v_1)$
$\quad = -5.75\,\text{m s}^{-1} - 1.5v_1$

so $2.5v_1 = -5.75\,\text{m s}^{-1}$ and $v_1 = -2.3\,\text{m s}^{-1}$ (to the left).
$v_2 = 3.5\,\text{m s}^{-1} + v_1 = 1.2\,\text{m s}^{-1}$ (to the right).

12

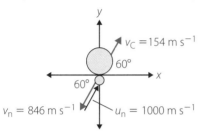

In the x direction:

(1) $u_{n,x} + v_{n,x} = u_{C,x} + v_{C,x}$ and

(2) $m_{n,x}u_{n,x} + m_{C,x}u_{C,x} = m_{n,x}v_{n,x} + m_{C,x}v_{C,x}$

From (1): $(1000\,\text{m s}^{-1})\cos 60° + v_{n,x} = 0\,\text{m s}^{-1} + v_{C,x}$
so $v_{C,x} = 500\,\text{m s}^{-1} + v_{n,x}$

From (2): $m(1000\,\text{m s}^{-1})\cos 60° + 0 = mv_{n,x} + 12mv_{C,x}$
so $v_{n,x} = (1000\,\text{m s}^{-1})\cos 60° - 12v_{C,x}$
$\quad = 500\,\text{m s}^{-1} - 12v_{C,x}$

Substitute the expression for $v_{C,x}$ from the equation derived from (1):

$v_{n,x} = 500\,\text{m s}^{-1} - 12v_{C,x}$
$\quad = 500\,\text{m s}^{-1} - 12(500\,\text{m s}^{-1} + v_{n,x}) = -5500\,\text{m s}^{-1} - 12v_{n,x}$

so $13v_{n,x} = 5500\,\text{m s}^{-1}$ and $v_{n,x} = -423\,\text{m s}^{-1}$ (the negative sign means it has bounced back to the left).

$v_{C,x} = 500\,\text{m s}^{-1} + v_{n,x} = 500\,\text{m s}^{-1} + (-423\,\text{m s}^{-1}) = 77\,\text{m s}^{-1}$

In the y direction:

(1) $u_{n,y} + v_{n,y} = u_{C,y} + v_{C,y}$ and (2)

$m_{n,y}u_{n,y} + m_{C,y}u_{C,y} = m_{n,y}v_{n,y} + m_{C,y}v_{C,y}$

From (1): $(1000\,\text{m s}^{-1})\sin 60° + v_{n,y} = 0\,\text{m s}^{-1} + v_{C,y}$
so $v_{C,y} = 866\,\text{m s}^{-1} + v_{n,y}$
From (2): $m(1000\,\text{m s}^{-1})\sin 60° + 0 = mv_{n,y} + 12mv_{C,y}$
so $v_{n,y} = (1000\,\text{m s}^{-1})\sin 60° - 12v_{C,y} = 866\,\text{m s}^{-1} - 12v_{C,y}$

Substitute the expression for $v_{C,y}$ from the equation derived from (1):

$v_{n,y} = 866\,\text{m s}^{-1} - 12v_{C,y}$
$\quad = 866\,\text{m s}^{-1} - 12(866\,\text{m s}^{-1} + v_{n,y}) = -9526\,\text{m s}^{-1} - 12v_{n,y}$

so $13v_{n,y} = -9526\,\text{m s}^{-1}$ and $v_{n,y} = -733\,\text{m s}^{-1}$ (the negative sign means it has bounced back downwards).

$v_{C,y} = 866\,\text{m s}^{-1} + v_{n,y} = 866\,\text{m s}^{-1} + (-733\,\text{m s}^{-1}) = 133\,\text{m s}^{-1}$

So

$v_C = \sqrt{v_{C,x}{}^2 + v_{C,y}{}^2} = \sqrt{(77\,\text{m s}^{-1})^2 + (133\,\text{m s}^{-1})^2} = 154\,\text{m s}^{-1}$

At angle to the horizontal given by

$\tan\theta = \dfrac{v_{C,y}}{v_{C,x}} = \dfrac{133\,\text{m s}^{-1}}{77\,\text{m s}^{-1}}, \theta = 60°.$

$v_n = \sqrt{v_{n,x}{}^2 + v_{n,y}{}^2} = \sqrt{(-423\,\text{m s}^{-1})^2 + (-733\,\text{m s}^{-1})^2}$
$\quad = 846\,\text{m s}^{-1}$

At angle to the horizontal given by

$\tan\theta = \dfrac{v_{n,y}}{v_{n,x}} = \dfrac{-733\,\text{m s}^{-1}}{-423\,\text{m s}^{-1}}, \theta = 60°.$

Note that we have done a lot of calculations here that we did not need to do! This is a silly way of doing this problem.

If we had simply changed our axes and set the x-axis to be along the direction of the velocity of the incoming neutron, we could have treated this as a one-dimensional problem and done it in just a few lines. Always look and think, before starting to do the maths!

1

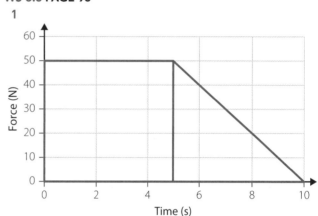

You can count the rectangles under the curve and multiply by the area of each rectangle, or divide the area into a rectangle and triangle as shown.

Rectangle area $=$ base $\times$ height
$\qquad\qquad\quad = 5\,\text{s} \times 50\,\text{N} = 250\,\text{N s} = 250\,\text{kg m s}^{-1}$

Triangle area $= \dfrac{1}{2} \times$ base $\times$ height
$\qquad\qquad = \dfrac{1}{2} \times 5\,\text{s} \times 50\,\text{N} = 250\,\text{N s} = 125\,\text{kg m s}^{-1}$

Total momentum transfer $= 250\,\text{kg m s}^{-1} + 125\,\text{kg m s}^{-1}$
$\qquad\qquad\qquad\qquad\;\; = 375\,\text{kg m s}^{-1}$

2 This question is easiest to do by estimating the number of rectangles below the curve and multiplying by the area of each rectangle. There are approximately 40 rectangles below the curve each with area $0.1\,\text{s} \times 20\,\text{N} = 2\,\text{N s} = 2\,\text{kg m s}^{-1}$, so the total momentum change is $80\,\text{kg m s}^{-1}$.

3 a

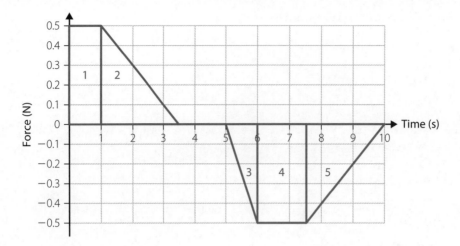

Divide the area into triangles and rectangles as shown.

Note the areas 3, 4 and 5 are all negative because the force is in the negative direction; that is, the momentum change is in the opposite direction from that for areas 1 and 2.

Areas 1 and 2 are positive, areas 3, 4, 5 are negative.

Area 1 = base × height = $1\,\text{s} \times 0.5\,\text{N} = 0.5\,\text{N s} = 0.5\,\text{kg m s}^{-1}$

Area 2 = $\frac{1}{2} \times$ base × height = $\frac{1}{2} \times 2.5\,\text{s} \times 0.5\,\text{N}$

$= 0.625\,\text{N s} = 0.625\,\text{kg m s}^{-1}$

Area 3 = $\frac{1}{2} \times$ base × height = $\frac{1}{2} \times 1\,\text{s} \times -0.5\,\text{N}$

$= -0.25\,\text{N s} = -0.25\,\text{kg m s}^{-1}$

Area 4 = base × height

$= 1.5\,\text{s} \times 0.5\,\text{N} = -0.75\,\text{N s} = -0.75\,\text{kg m s}^{-1}$

Area 5 = $\frac{1}{2} \times$ base × height

$= \frac{1}{2} \times 2.5\,\text{s} \times -0.5\,\text{N} = -0.625\,\text{N s} = -0.625\,\text{kg m s}^{-1}$

Total = $0.5\,\text{kg m s}^{-1} + 0.625\,\text{kg m s}^{-1} - 0.25\,\text{kg m s}^{-1}$
$- 0.75\,\text{kg m s}^{-1} - 0.625\,\text{kg m s}^{-1} = -0.5\,\text{kg m s}^{-1}$

The total momentum change is negative.

b $\vec{p}_\text{f} = \vec{p}_\text{i} + \Delta\vec{p} = \Delta\vec{p} = mv$

$v = \dfrac{\Delta\vec{p}}{m} = \dfrac{-0.5\,\text{kg m s}^{-1}}{0.1\,\text{kg}} = -5\,\text{m s}^{-1}$

4 a The solid line shows the force as a function of time.

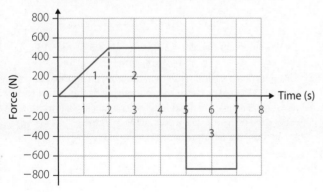

b Break up the area as shown by the rectangles and the triangle in part **a**.

Areas 1 and 2 are positive, area 3 is negative.

Area 1 = $\frac{1}{2} \times$ base × height = $\frac{1}{2} \times 2\,\text{s} \times 500\,\text{N} = 500\,\text{N s}$

$= 500\,\text{kg m s}^{-1}$

Area 2 = base × height = $2\,\text{s} \times 500\,\text{N} = 1000\,\text{N s} = 1000\,\text{kg m s}^{-1}$

Area 3 = $\frac{1}{2} \times$ base × height

$= 2\,\text{s} \times -750\,\text{N} = -1500\,\text{N s}$

$= -1500\,\text{kg m s}^{-1}$

Total = $500\,\text{kg m s}^{-1} + 1000\,\text{kg m s}^{-1} - 1500\,\text{kg m s}^{-1}$

$= 0\,\text{kg m s}^{-1}$

The total momentum change is zero.

c From part **b**, the momentum change after 2 s is $500\,\text{kg m s}^{-1}$. Because the car is starting from rest, the speed is

$v = \dfrac{\Delta\vec{p}}{m} = \dfrac{500\,\text{kg m s}^{-1}}{1200\,\text{kg}} = 0.42\,\text{m s}^{-1}$.

d From part **b**, using the areas of shapes 1 and 2, the momentum change after 5 s is $500\,\text{kg m s}^{-1} + 1000\,\text{kg m s}^{-1}$ $= 1500\,\text{kg m s}^{-1}$. Because the car is starting from rest, the

speed is $v = \dfrac{\Delta\vec{p}}{m} = \dfrac{1500\ \text{kg m s}^{-1}}{1200\ \text{kg}} = 1.25\,\text{m s}^{-1}$

e The car starts at rest and the total momentum change is zero, so it also finishes at rest.

....

WS 6.4 PAGE 96

1 Impulse, $\vec{I}$, is the change in momentum of an object when a force, $\vec{F}$, is applied to it over some time, t: $\vec{I} = \Delta\vec{p} = \vec{F}t$, with unit N s or kg m s^{-1}.

2 Impulse is the area under the curve of a force vs time graph. If you are studying calculus, this is the integral of force with respect to time.

3 $\Delta\vec{p} = \vec{F}t$, so on a graph of $\Delta\vec{p}$ vs t the gradient at any point is the force $\vec{F}$ at that time. If you are studying calculus, force is the derivative of momentum with respect to time.

4 $\vec{I} = \Delta\vec{p} = \vec{F}t$. David can increase the force he applies, or increase the time he applies it for: hit harder, or use a longer follow through.

5 a $\vec{I} = \Delta\vec{p} = \vec{F}t = 30\,\text{N} \times 0.1\,\text{s} = 3\,\text{N s} = 3\,\text{kg m s}^{-1}$

b $\vec{I} = \Delta\vec{p} = \vec{F}t$ so $t = \dfrac{\vec{I}}{\vec{F}} = \dfrac{10\,\text{kg m s}^{-1}}{30\,\text{N}} = 0.33\,\text{s}$ (noting that $1\,\text{N} = 1\,\text{kg m s}^{-2}$, so the units work)

c $\vec{I} = \Delta\vec{p} = \vec{F}t$ so $\vec{F} = \dfrac{\vec{I}}{t} = \dfrac{10\,\text{kg m s}^{-1}}{0.1\,\text{s}} = 100\,\text{kg m s}^{-2} = 100\,\text{N}$

6 a Convert to SI units:

$100\,\text{km h}^{-1} = 100\,\text{km h}^{-1} \times \dfrac{1000\,\text{m}}{1\,\text{km}} \times \dfrac{1\,\text{h}}{3600\,\text{s}}$

$= 27.7\,\text{m s}^{-1} = 28\,\text{m s}^{-1}$

$\vec{I} = \Delta\vec{p} = m\vec{v} = 1900\,\text{kg} \times 28\,\text{m s}^{-1} = 53\,000\,\text{kg m s}^{-1}$

b $\vec{I} = \Delta \vec{p} = \vec{F}t$ so

$$\vec{F} = \frac{\vec{I}}{t} = \frac{53\,000\,\text{kg m s}^{-1}}{12\,\text{s}} = 4400\,\text{kg m s}^{-2} = 4.4\,\text{kN}$$

c $\vec{I} = \Delta \vec{p}$, and the change in momentum is $53\,000\,\text{kg m s}^{-1}$ as the car comes to a stop.

d $\vec{I} = \Delta \vec{p} = \vec{F}t$ so

$$\vec{F} = \frac{\vec{I}}{t} = \frac{53\,000\,\text{kg m s}^{-1}}{0.75\,\text{s}} = 71\,000\,\text{kg m s}^{-2} = 71\,\text{kN}$$

e $\vec{I} = \Delta \vec{p} = \vec{F}t$ and $\Delta \vec{p} = m\Delta \vec{v}$, where m is now 65 kg and $\Delta \vec{v} = 28\,\text{m s}^{-1}$ because the passenger is initially at the same speed as the car, and their final speed must also be 0.

$$\vec{F} = \frac{\vec{I}}{t} = \frac{m\Delta \vec{v}}{t} = \frac{(65\,\text{kg})(28\,\text{m s}^{-1})}{0.75\,\text{s}} = 2400\,\text{kg m s}^{-2} = 2.4\,\text{kN}$$

f Crumple zones protect the people in a car during a crash by crumpling – increasing the time the collision takes, which reduces the forces because the total momentum change must be the same $\left(\Delta \vec{p} = \vec{F}\Delta t \right)$. The main function of the seatbelt is to stop people hitting the windscreen or other hard surfaces or other people in the car. But seatbelts are also slightly stretchy, so act to increase the time taken to slow the person down a bit. Seatbelts save a lot of lives and prevent a lot of injuries – always wear your seatbelt!

7

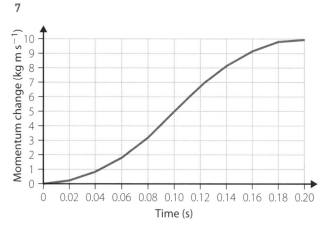

For $t < 0.1\,\text{s}$, the area under the curve is given by a triangle, so simply use area $= \frac{1}{2} \times$ base $\times$ height. This procedure gives:

t (s)	0	0.02	0.04	0.06	0.08	0.1
Area (N s)	0	0.2	0.8	1.8	3.2	5

For $t > 0.1\,\text{s}$, we can add the 5 N s for the first 0.1 s to the next slice of area along and use counting squares or some other method. Or, we can take the total area, which is

$\frac{1}{2} \times 0.2\,\text{s} \times 100\,\text{N} = 10\,\text{N s}$, and subtract the remaining area to the right. Either way, you get:

t (s)	0.12	0.14	0.16	0.18	0.2
Area (N s)	6.8	8.2	9.2	9.8	10

If you are studying calculus you can do this by using the fact that momentum is the integral of force with respect to time, and noting that for the first second $F = (1000\,\text{N s}^{-1})t$, which when integrated gives $p = (500\,\text{N s}^{-1})t^2$, and do this again for the second, where $F = (-1000\,\text{N s}^{-1})t$, which when integrated gives $p = (-500\,\text{N})t^2$, and remember to add the initial momentum from the first second.

Either way, you get the curve shown above, which is *not* a straight line, it is an s-curve.

8 a To the right

b To the left

c It's just to the left

WS 6.5 PAGE 99

1 In an elastic collision, changes in kinetic energy are small compared to the kinetic energy of the colliding objects, because internal friction forces are small. In a perfectly elastic collision there is no change in kinetic energy.

In an inelastic collision, kinetic energy is lost. In a perfectly inelastic collision the maximum possible kinetic energy is lost. This happens when the two objects stick together.

2 Scattering of subatomic particles can be perfectly elastic, as can gravity-assist manoeuvres, or any interaction that occurs due to field forces so there is no friction involved.

3 Anything where the two objects are stuck together afterwards, for example an insect hitting a windscreen, a meteorite hitting Earth.

4 a This is an inelastic collision, because the two objects (ball and Ellyse) are 'stuck together' after the collision.

b Using conservation of momentum and noting that the ball and cricketer move together with the same speed after the collision:

$$p_{\text{initial}} = m_{\text{ball}}u_{\text{ball}} + m_{\text{Ellyse}}u_{\text{Ellyse}}$$

$$= p_{\text{final}} = (m_{\text{ball}} + m_{\text{Ellyse}})v_{\text{ball+Ellyse}}$$

So $m_{\text{ball}}u_{\text{ball}} = (m_{\text{ball}} + m_{\text{Ellyse}})v_{\text{ball+Ellyse}}$

$$v_{\text{ball +Ellyse}} = \frac{m_{\text{ball}}u_{\text{ball}}}{m_{\text{ball}} + m_{\text{Ellyse}}} = \frac{(0.15\,\text{kg})(15\,\text{m s}^{-1})}{0.15\,\text{kg} + 65\,\text{kg}} = 0.035\,\text{m s}^{-1}$$

or $3.5\,\text{cm s}^{-1}$.

c Unless the ground is very slippery, the friction force from the ground will stop her moving completely. While we often ignore external forces when we analyse collisions, we need to always think about whether that is actually a reasonable approximation.

5 a We can calculate the velocity using conservation of momentum:

$p_i = p_f$, so $m_1u_1 + m_2u_2 = m_1v_1 + m_2v_2$
which, as they have the same mass, reduces to:
$u_1 + u_2 = v_1 + v_2$
$v_2 = u_1 + u_2 - v_1 = 6.5\,\text{m s}^{-1} + 4.5\,\text{m s}^{-1} - 5.0\,\text{m s}^{-1}$
$v_2 = 6.0\,\text{m s}^{-1}$

b If this was an elastic collision, the total kinetic energy would be the same before as after the collisions:

$$E_{\text{K, before}} = \frac{1}{2}m_1u_1^2 + \frac{1}{2}m_2u_2^2 = \frac{1}{2}m(u_1^2 + u_2^2)$$

$$= \frac{1}{2}m((6.5\,\text{m s}^{-1})^2 + (4.5\,\text{m s}^{-1})^2) = \frac{1}{2}m(62.5\,\text{m}^2\,\text{s}^{-2})$$

$$E_{\text{K, after}} = \frac{1}{2}m_1v_1^2 + \frac{1}{2}m_2v_2^2 = \frac{1}{2}m(v_1^2 + v_2^2)$$

$$= \frac{1}{2}m((5.0\,\text{m s}^{-1})^2 + (6.0\,\text{m s}^{-1})^2) = \frac{1}{2}m(61\,\text{m}^2\,\text{s}^{-2})$$

The total kinetic energy after the collision is less than that before, so this is an inelastic collision. However, it is not perfectly inelastic because the two balls are moving at different speeds after the collision.

6 a This is a perfectly inelastic collision – the clay is stuck to the floor afterwards.

   b There are several ways to do this. Using kinematics:
   $v^2 = u^2 + 2as$ where $u = 0$ and $a = g = 9.8\,\text{m s}^{-2}$:

   $$v = \sqrt{2as} = \sqrt{2(9.8\,\text{m s}^{-2})(1.3\,\text{m})} = 5.0\,\text{m s}^{-1}$$

   c The clay comes to a stop on colliding, so $\Delta \vec{p} = m\Delta \vec{v}$
   $= (0.5\,\text{kg})(5.0\,\text{m s}^{-1}) = 2.5\,\text{kg m s}^{-1}$ (up).

   d Momentum is conserved, so the change in Earth's momentum is the same, but in the opposite direction. $2.5\,\text{kg m s}^{-1}$, down.

   e $\Delta \vec{p} = m\Delta \vec{v}$ so $\Delta \vec{v} = \dfrac{\Delta \vec{p}}{m} = \dfrac{2.5\ \text{kg m s}^{-1}}{6.0 \times 10^{24}\ \text{kg}} = 4.2 \times 10^{-25}\,\text{m s}^{-1}$

   Which is approximately zero.

7 a Using the expression for kinetic energy: $E_k = \dfrac{1}{2}mv^2$ and rearranging:

   $$v = \sqrt{\frac{2E_k}{m}} = \sqrt{\frac{2(2.5 \times 10^{-14}\,\text{J})}{9.1 \times 10^{-31}\ \text{kg}}} = 2.3 \times 10^8\,\text{m s}^{-1}$$

   b $\vec{p} = m\vec{v} = (9.1 \times 10^{-31}\,\text{kg}\,(2.3 \times 10^8\,\text{m s}^{-1})$
   $= 2.1 \times 10^{-22}\,\text{kg m s}^{-1}$

   c Momentum is conserved, so the nitrogen nucleus recoils with the same momentum, but in the opposite direction. The original carbon nucleus had a mass of $2.3 \times 10^{-26}\,\text{kg}$ and this has decreased by $9.1 \times 10^{-31}\,\text{kg}$, so it is effectively unchanged given the precision we are using here.

   $$\vec{v} = \frac{\vec{p}}{m} = \frac{2.1 \times 10^{-22}\,\text{kg m s}^{-1}}{2.3 \times 10^{-26}\ \text{kg}} = 9300\,\text{m s}^{-1}\ \text{or}\ 9.3\,\text{km s}^{-1}$$

8 Call the initial direction of motion of the satellite x, then the initial direction of the meteorite is y.

   Using conservation of momentum in the two directions separately and then combining:

   In the x direction:
   $$\vec{p}_{\text{initial},x} = m_{\text{sat}}u_{\text{sat},x} + m_{\text{met}}u_{\text{met},x}$$

   $$= \vec{p}_{\text{final},x}(m_{\text{met}} + m_{\text{sat}})v_{\text{sat+met},x}$$

   $$v_{\text{sat+met},x} = \frac{m_{\text{sat}}u_{\text{sat},x}}{m_{\text{sat}} + m_{\text{met}}} = \frac{(3500\,\text{kg})(3100\,\text{m s}^{-1})}{3500\,\text{kg} + 250\,\text{kg}} = 2.9\,\text{km s}^{-1}$$

   In the y direction:
   $$\vec{p}_{\text{initial},y} = m_{\text{sat}}u_{\text{sat},y} + m_{\text{met}}u_{\text{met},y}$$

   $$= \vec{p}_{\text{final},y}(m_{\text{met}} + m_{\text{sat}})v_{\text{sat+met},y}$$

   $$v_{\text{sat+met},y} = \frac{m_{\text{met}}u_{\text{met},y}}{m_{\text{sat}} + m_{\text{met}}} = \frac{(250\,\text{kg})(15\,000\,\text{m s}^{-1})}{3500\,\text{kg} + 250\,\text{kg}} = 1.0\,\text{km s}^{-1}$$

   The total velocity is:

   $$v_{\text{sat+met}} = \sqrt{v_x^2 + v_y^2} = \sqrt{(2.9\,\text{km s}^{-1})^2 + (1.0\,\text{km s}^{-1})^2} = 3.1\,\text{km s}^{-1}$$

   The angle to the original velocity of the satellite is given by
   $$\tan\theta = \frac{v_y}{v_x} = \frac{1.0\,\text{km s}^{-1}}{2.9\,\text{km s}^{-1}}, \theta = 19°.$$

MODULE TWO: CHECKING UNDERSTANDING
PAGE 102

1 **C** must be directed south-west.

   Vector addition of a force directed south and a force directed west of equal magnitude will result in an isosceles right-angled triangle with internal angles of 45°. Consequently, the vector sum will be directly south-west.

2 **B** $T\cos 10°$

   Vector resolution of the tension force determines that the vertical component will be adjacent to the angle of 10° and this means its magnitude will be $T\cos 10°$.

3 **D** Tension (contact force), magnetic force (non-contact force) and gravitational force (non-contact force)

   As long as the string is taut there will be a contact tension force. A gravitational force will always act on objects on Earth and is a non-contact force. The question indicates that the object experiences a magnetic force that is also non-contact.

4 **A** The net force is zero but the tension force in each of the two wires is less than the weight force.

   The portrait is in equilibrium (it is not accelerating) and, therefore, must have net force of zero. Vector resolution of each tension force determines that the vertical component will be equal to $T\cos 20°$. There are two wires, each with a vertical component of $T\cos 20°$, so it is possible to say $2T\cos 20° = mg$. Rearranging we get an expression for the tension in each wire
   $$T = \frac{mg}{2\cos 20°}, \text{ which is less than } mg.$$

5 **C** The force that Earth exerts on the teacher is equal in magnitude to the force the teacher exerts on Earth but the teacher accelerates more because his mass is less.

   Newton's third law explains that for every action force there is an equal and opposite reaction force. Earth pulls down on the teacher; therefore, the teacher pulls up on Earth with a force of equal magnitude. Newton's second law relates the mass of the object to the acceleration that results from a force. Earth and the teacher experience equal and opposite forces but the teacher has a much smaller mass and thus accelerates more.

6 a The force required to lift an object at a constant speed is equal to the object's weight (mg).
   $W = Fs\cos\theta = (mg)s\cos\theta$
   $= (480 + 12 \times 65) \times 9.8 \times 40 \times \cos 0° = 493\,920\,\text{J} = 494\,\text{kJ}$

   b $\Delta U = mgh = 65 \times 9.8 \times 40 = 25\,480\,\text{J} = 25\,\text{kJ}$

   c $P = Fv = (mg) \times v = (480 + 12 \times 65) \times 9.8 \times 0.8 = 9878.4\,\text{W}$
   $= 10\,\text{kW}$

7 a Letting 'to the right' be positive:
   $\Sigma p_{\text{initial}} = 0.06 \times 0.75 + 0.04 \times -0.5 = 0.025\,\text{kg m s}^{-1}$
   Therefore, $\Sigma p_{\text{final}} = 0.025\,\text{kg m s}^{-1}$ (conservation of momentum)
   $\Sigma p_{\text{final}} = m_{\text{total}} \times v = 0.1 \times v$
   Therefore, $v = 0.25\,\text{m s}^{-1}$ or $0.25\,\text{m s}^{-1}$ to the right (since it is positive).

   b $\Delta p = m\Delta v = m \times (v - u) = 0.04 \times (0.25 - -0.5) = 0.03\,\text{kg m s}^{-1}$ to the right (since it is positive).

   c An elastic collision is one in which kinetic energy is conserved, that is $\Sigma K_{\text{initial}} = \Sigma K_{\text{final}}$.

   $$\Sigma K_{\text{initial}} = \frac{1}{2}m_A v_A{}^2 + \frac{1}{2}m_B v_B{}^2$$

   $$= \frac{1}{2} \times 0.04 \times (-0.5)^2 + \frac{1}{2} \times 0.06 \times (0.75)^2 = 0.022\,\text{J}$$

   $$\Sigma K_{\text{final}} = \frac{1}{2}m_{A+B}v^2 = \frac{1}{2} \times 0.10 \times (0.25)^2 = 0.003\,\text{J}$$

   Since $\Sigma K_{\text{initial}}$ does not equal ΣK_{final} the collision is not elastic.

   d $\Delta p_A = F_{\text{B on A}}\Delta t = 0.03\,\text{kg m s}^{-1}$ (from part **b**)

   Therefore, $F_{\text{B on A}} = \dfrac{\Delta p}{\Delta t} = \dfrac{0.03}{0.2} = 0.15\,\text{N right}.$

   This will be equal and opposite to the force that ball A exerts on ball B (Newton's third law).

8 a The block moves at a constant speed so there is no net force acting; that is, the forces acting on it are balanced. The force down the ramp is balanced by the force of friction and the force into the ramp is balanced by the normal force. Hence the respective vector arrows are drawn to the same length.

b $\sin\theta = \dfrac{F_{g\text{ down slope}}}{F_g}$

Therefore, $F_{g\text{ down slope}} = F_g \times \sin\theta$

$F_{g\text{ down slope}} = mg \times \sin\theta = 5g\sin 8°$

c $\cos\theta = \dfrac{F_{g\text{ into slope}}}{F_g}$

Therefore, $F_{g\text{ into slope}} = F_g \times \cos\theta$

$F_{g\text{ into slope}} = mg \times \cos\theta = 5g\cos 8°$

d $F_{friction} = \mu \times F_{normal}$

And $F_{normal} = F_{g\text{ into slope}}$

Therefore, $F_{friction} = \mu \times F_{g\text{ into slope}} = \mu \times 5g\cos 8°$ (from **c**)

e $F_{friction} = F_{g\text{ down slope}} = 5g\sin 8°$ (from **b**)

And $F_{friction} = \mu \times F_{g\text{ into slope}} = \mu \times 5g\cos 8°$ (from **d**)

Therefore, $F_{g\text{ down slope}} = \mu \times F_{g\text{ into slope}}$ (since both equal $F_{friction}$)

Therefore, $5g\sin 8° = \mu \times 5g\cos 8°$

$\mu = \dfrac{5g\sin 8°}{5g\cos 8°} = \tan 8° = 0.14$

MODULE THREE: WAVES AND THERMODYNAMICS

REVIEWING PRIOR KNOWLEDGE **PAGE 105**

1 A wave is the transfer of energy with no net movement of matter/mass.

2

A	Conduction	2	1	Heat coming from the Sun
B	Convection	3	2	A metal pan heating up on a stove
C	Radiation	1	3	Warm air circulating in a room

3 a The distance between two consecutive peaks or troughs of the wave.

  b The number of times a wave passes a specific point.

  c The time take for a wave to cover a certain distance.

4 a Transverse wave

  b Longitudinal wave

5 Refraction

6 Reflection

7 Answers will vary but as long as they are found on the electromagnetic spectrum in the correct order then the answer is acceptable.

Chapter 7: Wave characteristics

WS 7.1 **PAGE 106**

2

What are the risks in doing this experiment?	How can you manage these risks to stay safe?
String wrapped around the fingers	If the string is too tight or is left on too long, it may restrict blood flow to the fingers. Make sure that the string is not too tight and is removed from fingers between trials.
Putting fingers in your ears	If the fingers are placed too far into the ear it may cause damage, so don't push fingers too far into the ear.
May drop the oven rack	Ensure that the rack is over something soft and not too far off the ground

3 1 Fill a suitable container with 2 cm of water and place on a flat surface.

  2 Set up a recording device to be able to see the creation of the waves.

  3 Place a piece of floating material onto the water.

  4 Push the material up and down and record the resulting direction of the waves.

  5 Move the material backwards and forwards in the water and record the direction of the waves.

  6 Watch the video to check your observations.

5 a Answers should include some discussion on the effect of the medium, such as speed of the waves and spread of energy.

  b Answer if they could change direction of the waves in the mediums.

  c Answer how they would improve the investigation, such as repetition or more accurate recording.

6 Answers will vary, but students need to have related their conclusion to their hypothesis.

7 Answers will vary, but should be a testable statement such as 'by moving one jelly baby we will see the energy transfer along the wave'.

8 Risks can include: being stabbed with a skewer – make sure the points of the skewer are away from people; slipping on a jelly baby – ensure that all waste jelly babies are picked up.

9 Answers should include that as the skewer was lifted it caused other skewers to be raised in a direction of travel away from the first skewer. The difference between transverse and longitudinal waves is that in longitudinal waves in the movement of the particles is in the same direction as the direction of transfer of energy; in transverse waves it is at right angles.

10 Answers will vary, but should relate back to the hypothesis.

11 How does the direction of the transfer of energy compared to the movement of the particles change the name of the wave?

12 Answers will vary, but should be relevant to the investigation.

13 Answers will vary, but should make note of the relationship of the movement of the ribbon to the direction of transfer of energy.

14 Answers will vary, but should include observations that the ribbon moved around one spot but the energy moved along the slinky, as well as the direction of movement of the ribbon in relation to the direction of movement of the energy.

15 Answers will vary, but should always relate to the hypothesis.

16 An electromagnetic wave does not require a medium because the interaction between the oscillating charge transferring the energy to the electric field around the particle causes a change in the magnetic field and allows the propagation of energy through a vacuum.

17 The film *Alien* is set in space, and because space is a vacuum then there is no medium for a wave to travel through. Sound is a mechanical wave so it requires a medium for the energy to travel through, as the energy is passed through the oscillations of a particle around a central point passing on the kinetic energy by collisions with the particle next to it. There are no particles in a vacuum so it would be impossible for a mechanical wave to be propagated through it. Therefore, the tagline is true unless they were on a space ship with its own atmosphere.

18

Type of wave	Wavelength	Frequency (Hz)	Use
Gamma	1×10^{-12} m and below	10^{20}–10^{24}	To kill cancer cells, sterilise medical equipment
X-ray	1 nm – 1 pm	10^{17}–10^{20}	For imaging purposes

Type of wave	Wavelength	Frequency (Hz)	Use
Ultraviolet	400 nm – 1 nm	$10^{15} - 10^{17}$	Kill microbes and sterilise equipment
Visible light	Range from 700–400 nm	$4 - 7.5 \times 10^{14}$	To see things with the naked eye
Infrared	750 nm – 25 µm	$10^{13} - 10^{14}$	Thermal imaging cameras
Microwave	1 mm – 25 µm	$3 \times 10^{11} - 10^{13}$	Cooking food, communication
Radio wave	Greater than 1 mm	Greater than 3×10^{11}	Communication

WS 7.2 PAGE 113

1

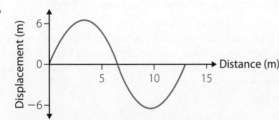

2 As there are four divisions and each is 4 ms, the period would be 16 ms.

3 a i 10 s
  ii 0.1 Hz
  ii 4 cm
  b i 40 s
  ii 0.025 Hz
  iii 6 cm
  c i 20 s
  ii 0.05 Hz
  iii 2.5 cm
  d i 0.6 s
  ii 1.7 Hz
  iii 6 cm

4

5 a 35 cm
  b 30 Pa to –30 Pa
  c It is a longitudinal wave because the particles are oscillating parallel to the direction of energy transfer. It is also a sound wave, which is by its nature a longitudinal wave.

WS 7.3 PAGE 117

1 The period of the wave is 0.05 s and $f = \frac{1}{T}$. Therefore
$$f = \frac{1}{0.05 \text{ s}} = 20 \text{ Hz}$$

2 The period of the wave is the inverse of the frequency, so
$$T = \frac{1}{f} \therefore T = \frac{1}{100 \text{ Hz}} = 0.01 \text{ s}.$$

3 Frequency is a rate quantity that refers to the number of cycles per second, and the period is a time quantity that refers to the number of seconds per cycle. We have an inverse relationship such that when the frequency is high the period is low and vice versa.

4 To find the velocity of the wave we need to use $v = f\lambda$.
Period = 3 s, $\lambda = 1.4$ m
$$f = \frac{1}{T} \therefore \frac{1}{3 \text{ s}} = 0.33 \text{ Hz}$$
Substituting this into $v = f\lambda$ gives $v = 0.33 \text{ Hz} \times 1.4 \text{ m} = 4.7 \text{ m s}^{-1}$.

5

Type of wave	λ	Frequency
Radio	13 m	2.3×10^7 Hz
Microwave	9.23×10^{-3} m	3.25×10^{10} Hz
Red light	680 nm	4.41×10^{14} Hz
Blue light	454 nm	660 THz
X-ray	0.350 pm	8.57×10^{20} Hz
Gamma ray	6.52×10^{-12} m	4.6×10^{19} Hz

6 a If the speed of sound is 340 m s^{-1} and it takes 3 s for the sound to reach you, the distance would be $3 \times 340 \text{ m s}^{-1} = 1020 \text{ m}$ away.

  b As the distance is 1020 m we can use $t = \frac{s}{v}$.
$$\therefore \frac{1020 \text{ m}}{3 \times 10^8 \text{ m s}^{-1}} = 3.4 \times 10^{-6} \text{ s}$$

7 a Nine waves per minute so the period of the wave is
$$\frac{60}{9} = 6.7 \text{ s}.$$
  b $f = \frac{1}{T} \therefore \frac{1}{6.7 \text{ s}} = 0.15 \text{ Hz}$
  c Using $v = f\lambda$, $v = 0.15 \text{ Hz} \times 15 \text{ m} = 2.25 \text{ m s}^{-1}$.

8 a Wavelength is given by $f = \frac{1}{T}$ and $v = f\lambda$
$$\therefore \text{ in air, } f = \frac{1}{7.0 \times 10^{-3} \text{ s}} = 143 \text{ Hz}$$
Substituting in $v = f\lambda$, $\lambda = \frac{v}{f} = \frac{340 \text{ m s}^{-1}}{143 \text{ Hz}} = 2.4 \text{ m}$
In the copper rod, $\lambda = \frac{v}{f} = \frac{3750 \text{ m s}^{-1}}{143 \text{ Hz}} = 26 \text{ m}$

  b The velocity changes due to the different state of the material. Sound waves travel faster through solids than they do through gas because in solids the particles are closer to each other. The frequency stays the same because the period of the wave that has caused the wave in the first place remains constant, therefore from $v = f\lambda$ the wavelength will increase with a greater velocity.

9 A radio wave is an electromagnetic wave and therefore travels at the speed of light $(3 \times 10^8 \text{ m s}^{-1})$. Using $v = f\lambda$,
$$\lambda = \frac{v}{f} = \frac{3 \times 10^8 \text{ m s}^{-1}}{1.017 \times 10^8 \text{ Hz}} = 2.9 \text{ m}.$$

Chapter 8: Wave behaviour

WS 8.1 PAGE 119

1 The angle of reflection of the wave is the same as its angle of incidence. This applies to all points along the surface. The angles are measured with respect to the normal.

2

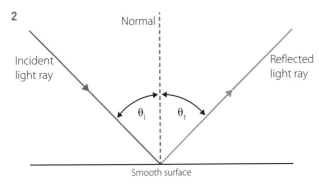

Normal

Incident light ray

Reflected light ray

θ_i θ_r

Smooth surface

3 A satellite dish uses a concave surface to reflect the received rays towards a central antenna. It does this to provide a larger surface area on which the required rays can be received and collected. This mechanism is also used in reflecting light and radio telescopes, to increase the amount of light, making the picture clearer.

4 A convex mirror gives the impression of a fisheye view, providing a greater field of vision at a smaller magnification. This means convex mirrors are good for showing oncoming traffic in a carpark.

5

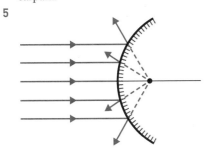

The image produced by a convex mirror is always virtual and located behind the mirror. When the object is far away from the mirror, the image is upright and located at the focal point. As the object approaches the mirror the image also approaches the mirror, and grows until its height equals that of the object.

6

What are the risks in doing this investigation?	How can you manage these risks to stay safe?
Hurt yourself on the butter knife	Hold the knife by the handle and don't run with it

7

	Concave spoon	Convex spoon	Flat surface
Image from far away	Reduced and inverted	Reduced and the right way up	Reduced and the correct way up left right reversal
Image up close	Enlarged and the right way up	Reduced and the right way up	Enlarged and the right way up left right reversal

8 From the convex side of the spoon, the rays diverge.

From the concave side of the spoon, the rays converge.

In the flat knife, the reflected rays are parallel to each other.

9 The concave shape would cause the rays to be focused into a point in front of the surface.

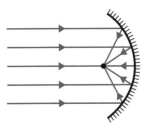

The convex shape would cause the rays to rays diverge from a focal point behind the surface and thus a virtual image is formed.

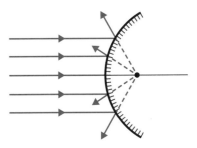

The flat surface would cause the rays to reflect back at the same angle that they hit the surface:

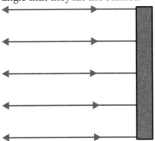

10 The image in the concave side of the spoon is inverted. As the object moves closer to the spoon, the image becomes larger.

The image on the convex side and the flat surface is upright. It also gets larger as the object approaches the mirror.

11 If the wave is travelling to a denser medium such as air to water, as the wave hits the surface of the new medium it will slow down and the wavelength will shorten. The opposite will be true going the other way, from a denser to less dense medium. This will have the effect of causing the wave front to bend towards the normal as the far side of the wave is traveling faster than the near side, causing the wave front to turn.

12 a Warm air (1)

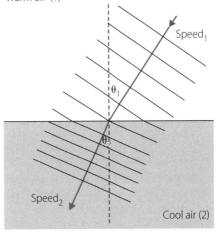

Speed$_1$

θ_1

Speed$_2$

Cool air (2)

b

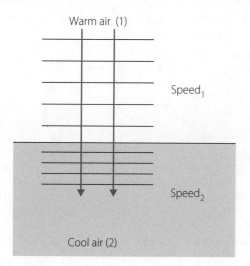

Warm air (1)

Speed₁

Cool air (2)

$Speed_1$

$Speed_2$

13 In a swimming pool, objects within the pool appear to be closer to the observer standing outside of the pool, due to the light being refracted at the surface.

In glasses, the light is being refracted to allow the lens in the eye to focus to produce a clearer image. The type of lens required will depend on whether the person suffers from near or far sightedness.

14 shorter, less, longer

15

a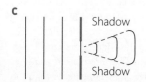
Shadow

d

b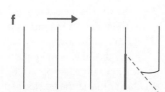

e
Shadow

c
Shadow
Shadow

f

16 The principle of superposition may be applied to waves when two or more waves travel through the same medium at the same time. The waves pass through each other without being disturbed. The net displacement of the medium at any point in space or time is the sum of the individual wave displacements.

17 Constructive interference is when two or more waves of the same frequency and in phase interact, producing a resultant wave which is the sum of the individual waves.

Destructive interference is when two waves of the same frequency but opposite phase interact to cancel each other out. This happens when the positive displacement of one wave interacts with the negative displacement of the other.

18 Constructive interference can be seen on a guitar string. When the vibrations travel back and forth they will constructively interfere and produce a standing wave.

Destructive interference can be seen in noise-cancelling headphones. They work by a microphone in the headphones picking up an incoming wave and sending out a wave which is the opposite so the user cannot hear noise from the outside world.

19 a, b, c

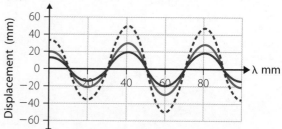

20

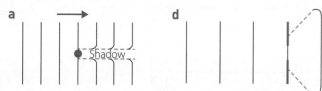

wave c = wave a + wave b

21

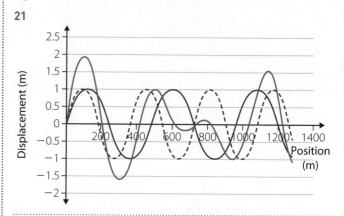

WS 8.2 PAGE 125

1 A standing wave has no net flow of energy. It is caused by constructive interference on a fixed string or pipe where two waves of the same wavelength reflect past each other. They are characterised by having points that appear not to move (nodes) and points that move through a maximum distance (antinode). In contrast, a progressive wave will have peaks and troughs that pass through the medium and has a net energy flow.

2

Property	Standing wave	Progressive wave
Frequency	All particles oscillate at same frequency except those at the nodes	All particles oscillate at same frequency
Amplitude of oscillations	Zero at nodes, maximum at antinodes	The same along the wave
Amplitude of wave	Doubles at constructive interference points, cancels at destructive interference points	The same along the wave

3 The video should show that the waves in the tube are moving forwards and backwards until the wave appears to be standing still.

4 The progressive wave has the transfer of energy moving from one point to another all along the tube. The standing wave has parts of the wave with maximum displacement (antinode) and parts of the wave where there is no movement of the particles (node).

5 The frequency is determined by the source of the wave, so this will remain constant. The velocity of the wave may increase, which would see a decrease in the resulting wavelength. This can be seen in the relationship $f = \dfrac{v}{\lambda}$.

9780170449595

WS 8.3 PAGE 127

1 The increase in the amplitude of oscillation of the particles in a system when they are exposed to a periodic force which is very close to the natural frequency of the system.

2 Students can include several different items including a guitar string, a column of soldiers marching in time, a swing, the Tacoma Narrows bridge, the Millennium footbridge.

3 Depending on what the student has chosen, they need to include the facts that the waves are being amplified by the natural frequency being met by the driving frequency. As well, when the driving frequency is matched the maximum energy is transferred between the two systems. They should also include the outcome of the resonance as well as any issues that this may have caused.

Chapter 9: Sound waves

WS 9.1 PAGE 128

1 As a wave increases in pitch the wavelength will shorten, but no change will be seen in the frequency when the loudness is increased.

2

What are the risks in doing this investigation?	How can you manage these risks to stay safe?
Playing sounds at loud volumes can damage hearing	Don't stand too close to the signal generator or play for too long
Can drop the tuning fork onto a person	Make sure that the tuning fork is held securely.
Using unsafe electrical appliances can lead to electric shock	When using electrical appliances, ensure that they are tested and used appropriately

3 1 Download the apps.
   2 Open the oscilloscope on either your phone or a computer.
   3 Set the signal generator to 550 Hz.
   4 Pause the wave generated in the oscilloscope and record the details of the wave, such as the amplitude and the wavelength. From this you can work out the period of the wave.
   5 Change the signal frequency and repeat the preceding steps.
   6 Turn the volume up on the signal generator and record the wave characteristics.

4 As the frequency increased the wavelength got shorter and the pitch got higher. As the loudness of the sound increased the amplitude of the wave increased. Changing the volume did not alter the frequency or the wavelength. As the wavelength shortened the period of the wave decreased, and vice versa.

5 The pitch of a sound is determined by its frequency. Increasing the frequency results in an increase in pitch and decrease in wavelength. The loudness/volume of a sound is directly related to its amplitude. An increase in amplitude causes an increase in volume.

WS 9.2 PAGE 130

1 a and b

	C	R
Meaning of letter	compression	rarefraction
Pressure level	high	low

2 1 Set up a video camera to record the experiment.
   2 Have one person stand and hold the slinky.
   3 Have another person hold the other end of the slinky and move 4–5 m away from them.
   4 One person holds the slinky still while the other is to move the slinky up and down to form a wave.
   5 Once this has been recorded, stretch the slinky out so that it is taut and bunch up several of the coils on the slinky. Let go of the coil and record what happens to the movement of the slinky.

3 As the particles oscillate around a point they knocked into other particles. This in turn caused those particles to oscillate around a point, showing the transfer of energy down through the particles without them moving.

4 As seen with mechanical waves, the particles pass on energy through movement. The sound energy causes the first particles to start to vibrate and then pass on the kinetic energy. Therefore, sound needs to have particles to pass on the energy, making it a mechanical wave.

5 As the particles compress they come closer together. This will lead to areas of higher pressure in the system. In the areas of rarefaction the particles are further apart, and therefore the pressure is less in these areas.

WS 9.3 PAGE 132

1 As with other sources of energy that emanate from a point source, sound will be subject to the inverse square law and as such will lose intensity in proportion to the square of distance.

2 $I_1 = 3.00 \times 10^{-7}\,\text{W m}^{-2}, I_2 = ?, r_1 = 2.00\,\text{m}, r_2 = 5.00\,\text{m}$

$$I \propto \frac{1}{r^2}$$

$$\frac{I_1}{I_2} = \left(\frac{r_2}{r_1}\right)^2$$

$$I_1 r_1^2 = I_2 r_2^2$$

$$I_2 = \frac{I_1 r_1^2}{r_2^2} = \frac{(3.00 \times 10^{-7}\,\text{W m}^{-2}) \times (2.00\,\text{m})^2}{(5.00\,\text{m})^2}$$

$$= 4.80 \times 10^{-8}\,\text{W m}^{-2}$$

3 a $I_1 = 3.20\,\text{W m}^{-2},\ I_2 = 8.00 \times 10^{-4}\,\text{W m}^{-2},$
   $r_1 = 15\,\text{m}, r_2 = ?$

$$I \propto \frac{1}{r^2}$$

$$\frac{I_1}{I_2} = \left(\frac{r_2}{r_1}\right)^2$$

$$I_1 r_1^2 = I_2 r_2^2$$

$$r_2^2 = \frac{I_1 r_1^2}{I_2}$$

$$r_2 = \sqrt{\frac{I_1 r_1^2}{I_2}} = r_1\sqrt{\frac{I_1}{I_2}} = 15\sqrt{\frac{3.20}{8.00 \times 10^{-4}}}$$

$$r_2 = 948.68\,\text{m} \approx 950.00\,\text{m}$$

b $I \propto \dfrac{1}{r^2}$ so

$$15\%\ I = 0.15 = \frac{1}{r^2}$$

$$r^2 = \frac{1}{0.15} = 6.67$$

$$r = \sqrt{6.67} = 2.58\,\text{m} \approx 2.6\,\text{m}$$

4 $I_1 = 2.00 \times 10^{-3}\,\mathrm{W\,m^{-2}}$, $I_2 = ?$, $r_1 = 60.0\,\mathrm{m}$, $r_2 = 1.00\,\mathrm{m}$ (from where the operator's head is from the jackhammer)

$$I \propto \frac{1}{r^2}$$

$$\frac{I_1}{I_2} = \left(\frac{r_2}{r_1}\right)^2$$

$$I_1 r_1^2 = I_2 r_2^2$$

$$I_2 = \frac{I_1 r_1^2}{r_2^2} = \frac{2.00 \times 10^{-2} \times 60.0^2}{1.00^2}$$

$$I_2 = 7.2\,\mathrm{W\,m^{-2}}$$

The operator would not lose hearing immediately because that would require an intensity of over $10\,\mathrm{W\,m^{-2}}$. However, prolonged exposure of sound intensity at this level would cause hearing damage over time.

WS 9.4 PAGE 134

1 Echolocation. By sending out a sound wave and receiving the echo back at a given time, the distance can be worked out.

$\text{Speed} = \dfrac{\text{distance}}{\text{time}}$; distance = speed × time. This allows for an organism or machine to work out depth and map an area using sound.

2

Equipment used	Observations
Rubber band across your fingers	The rubber band made a 'twanging' sound and the more it was stretched the higher the pitch

~continued in right column ▲

Equipment used	Observations
Rubber band across an ice-cream container	The sound was similar to across the fingers but was louder due to being amplified by the ice-cream container
Rubber bands across an ice-cream container with cloth in it	The sound was dampened by the cloth in the ice-cream container

3 In the first situation the vibrations of the rubber band caused the energy to dissipate to the surrounding environment so it could not be heard easily. When the rubber band was stretched over the container, the sound waves were able to reflect off the container and constructively interfere with each other to give a louder sound than the rubber band by itself. When the cloth was put into the container it caused the sound to reflect at lots of different angles, as well as being absorbed by the softer material. This had the effect of damping the sound as the waves were not directed back to the listener because the waves were not able to interfere constructively.

4 The sound reflected off a hard surface gave an overall amplified sound due to the waves constructively interfering with each other. Softer materials prevented the sound reflecting.

WS 9.5 PAGE 136

1 When you send vibrations down a string or a pipe it will cause them to reflect off the end point. This will then cause destructive or constructive interference. When we have constructive interference we create harmonics, which is what we know as notes.

2 The areas of maximum movement are called antinodes and are caused by constructive interference. The areas of minimum/ no movement are called node and are caused by destructive interference.

3

Vibration mode	Wave pattern	f and λ (in terms of v and l)
Fundamental mode of vibration 1st harmonic		$\lambda_1 = 2l$ $f_1 = \dfrac{v}{2l}$
1st overtone 2nd harmonic		$\lambda_2 = l$ $f_2 = 2f_1$
2nd overtone 3rd harmonic		$\lambda_3 = \dfrac{2l}{3}$ $f_3 = 3f_1$
3rd overtone 4th harmonic		$\lambda_4 = \dfrac{2l}{4} = \dfrac{l}{2}$ $f_4 = 4f_1$

9780170449595

4

Vibration mode	Particle displacement	Pressure variation	f and λ
Fundamental mode of vibration 1st harmonic	l Antinode　　　Antinode	Node　　　　　Node	$\lambda_1 = 2l$ $f_1 = \dfrac{v}{2l}$
1st overtone 2nd harmonic			$\lambda_2 = l$ $f_2 = 2f_1$
2nd overtone 3rd harmonic			$\lambda_3 = \dfrac{2l}{3}$ $f_3 = 3f_1$
3rd overtone 4th harmonic			$\lambda_4 = \dfrac{2l}{4} = \dfrac{l}{2}$ $f_4 = 4f_1$

5 **a** You would weigh the mass carrier and the individual masses twice and record the resulting mass. You can then work out the average uncertainty by dividing the range of the measurements by 2.

b Measure the wavelength in three different places where you can still clearly see one wavelength. This should be at resonance, maximum wavelength and minimum wavelength. Work out uncertainty by using the range between maximum and minimum wavelengths measured.

6 **a** 300 g including the 50 g mass of the carrier

b Independent variable is the mass and dependent variable is the wavelength (λ (cm)).

7

Total mass on the end of the string (g)	Tension force (N)	$\dfrac{\lambda}{2}$ (cm)	λ (cm)	$\dfrac{3\lambda}{2}$ (cm)	Average λ (cm)
50 ± 1.25	0.49 ± 0.01225	15	30	45	30
100 ± 2.5	0.98 ± 0.0245	21	42	63	42
150 ± 3.75	1.47 ± 0.03675	25.25	50.5	73.75	49.83
200 ± 5	1.96 ± 0.049	28.5	57	85.5	57
250 ± 6.25	2.45 ± 0.06125	31.75	63.5	95.25	63.5
300 ± 7.5	2.94 ± 0.0735	34.75	69.5	104.25	69.5

8 2.5%

9 **a**

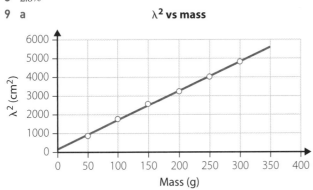

b As λ^2 is the dependent variable and m is the independent variable, it leaves the slope $= \dfrac{g}{f^2\mu}$.

c To find μ, $\mu = \dfrac{g}{\text{slope} \times f^2} = \dfrac{9.8}{17.28 \times 50^2} = 2.26\,\mathrm{g\,m^{-1}}$.
Answers may vary, depending on how the graph is plotted.

d It enables us to linearise the data and to work out the correct density of the piece of string.

WS 9.6 PAGE 140

1 **a** $v = 340\,\mathrm{m\,s^{-1}}$, $v_{\text{source}} = 60\,\mathrm{km\,h^{-1}}$.

$$60\,\mathrm{km\,h^{-1}} = \frac{60\,\mathrm{km} \times 1000\,\mathrm{m\,km^{-1}}}{1\,\mathrm{h} \times (60 \times 60)\,\mathrm{s\,h^{-1}}} = \left(\frac{60}{3.6}\right)\mathrm{m\,s^{-1}} = 16.67\,\mathrm{m\,s^{-1}}$$

$f = 1450\,\mathrm{Hz}$, $f' = ?$

$$f' = \left(\frac{v}{v - v_s}\right)f$$

$$f' = \frac{340}{340 - 16.67} \times 1450$$

$$f' = 1524.74 \approx 1525\,\mathrm{Hz}$$

b $v = 340\,\mathrm{m\,s^{-1}}$, $v_{\text{source}} = 16.67\,\mathrm{m\,s^{-1}}$, $f = 1450\,\mathrm{Hz}$, $f' = ?$

$f' = \left(\dfrac{v}{v + v_s}\right)f$. As the source is moving away from the observer, v_s is negative because it is a vector. This gives the equation

$$f' = \left(\frac{v}{v - (-v_s)}\right)f = \left(\frac{v}{v + v_s}\right)f$$

$$f' = \frac{340}{340 + 16.67} \times 1450$$

$$f' = 1382.23 \approx 1382\,\text{Hz}$$

2 a Because Serena is the one moving in relation to Rocco, her horn would be the source and Rocco would be the observer. As before, to convert km h^{-1} to m s^{-1} we divide by 3.6.

$$f' = ?, v_\text{wave} = 340\,\text{m s}^{-1}, v_\text{source} = 30.56\,\text{m s}^{-1},$$

$$v_\text{observer} = 19.44\,\text{m s}^{-1}, f = 182\,\text{Hz}$$

$$f' = f\left(\frac{v_\text{wave} + v_\text{observer}}{v_\text{wave} - v_\text{source}}\right) = 182 \times \left(\frac{340 + 19.44}{340 - 30.56}\right)$$

$$f' = 211\,\text{Hz}$$

b Because we have changed who is moving compared to whom, we just switch around the observer and source.

~continued in right column ▲

$$f' = ?, v_\text{wave} = 340\,\text{m s}^{-1}, v_\text{source} = 19.44\,\text{m s}^{-1}$$

$$v_\text{observer} = 30.56\,\text{m s}^{-1}, f = 182\,\text{Hz}$$

$$f' = f\left(\frac{v_\text{wave} + v_\text{observer}}{v_\text{wave} - v_\text{source}}\right) = 182 \times \left(\frac{340 + 30.56}{340 - 19.44}\right)$$

$$f' = 210\,\text{Hz}$$

3 Using $f_1 = 440\,\text{Hz}$, $f_2 = 431\,\text{Hz}$

$$f_\text{beat} = |f_2 - f_1| = |431 - 440|$$

$$f_\text{beat} = 9\,\text{Hz}$$

4 Using $f_\text{beat} = 7\,\text{Hz}$ and $f_1 = 390\,\text{Hz}$

$$f_\text{beat} = |f_2 - f_1|$$

$$f_2 = |f_1 - f_\text{beat}| = |390 - 7|$$

$$f_2 = 383\,\text{Hz or }397\,\text{Hz}.$$

Chapter 10: Ray model of light

WS 10.1 PAGE 142

1

Converging lenses (thicker in the centre)

 a Bi-convex

 b Plano-convex

 c Converging meniscus

 d Concave spherical

 e Concave parabolic

k Plane mirror

Diverging lenses (thinner in the centre)

 f Bi-concave

 g Plano-concave

h Diverging meniscus

 i Convex spherical

 j Convex parabolic

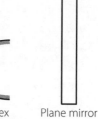 Plane mirror

2

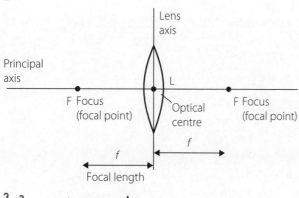

3 a

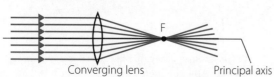

b

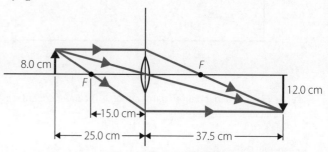

4 a

b The image is real because it is inverted and the actual light rays cross each other; therefore they can be projected onto a screen.

c 12 cm, so it is enlarged.

9780170449595

5 a

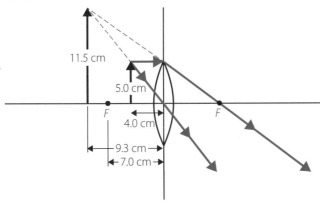

11.5 cm
5.0 cm
4.0 cm
F
9.3 cm
7.0 cm
F

b The image appears to be on the same side of the object and is the right way up. This would indicate that the rays of light do not cross but only appear to, making the image virtual.

c The image will be approximately 11.5 cm tall if the diagram is drawn accurately.

6 a

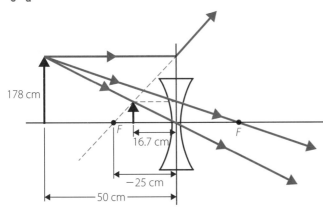

178 cm
F
16.7 cm
−25 cm
50 cm
F

b The image as seen in the diagram is a virtual image because it is upright and the rays of light do not cross but only appear to. It is also a smaller image.

c From the diagram drawn the image position should be 16.7 cm on the same side as Sebastian.

d From the diagram drawn, students should be able to show the image is approximately 59.3 cm tall.

7 a

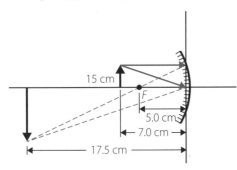

15 cm
F
5.0 cm
7.0 cm
17.5 cm

b From the drawing we can see that it is a real image because it is inverted and the rays of light cross.

c Using the diagram, the students should be able to show that the image is approximately 17.5 cm in front of the mirror.

d From the diagram the students should be able to tell that the image is approximately 37.5 cm tall.

8 a From the diagram, students should have the image distance approximately 47.7 cm behind the mirror.

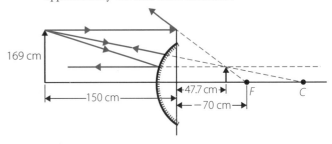

169 cm
150 cm
47.7 cm
70 cm
F
C

b The rays do not cross so it would be a virtual image that is upright and diminished.

c From the diagram the image should be approximately 53.7 cm.

9 $u = 25\,cm, f = 15\,cm, v = ?$

$$\frac{1}{u} + \frac{1}{v} = \frac{1}{f} \quad \text{so} \quad \frac{1}{v} = \frac{1}{u} - \frac{1}{f} = \frac{u-f}{fu}$$

$$v = \frac{fu}{u-f} = \frac{15.0 \times 25.0}{25.0 - 15.0} = 37.5\,cm$$

This shows that the image is on the opposite side of the lens and is at the point where the rays cross, making it a real image.

10 Using $u = 25\,cm$ and $v = 37.5\,cm$, $M = -\frac{v}{u} = \frac{37.5}{25} = 1.5$.

Using $M = \frac{h_i}{h_o}$:

$$h_i = Mh_o = 8.00 \times 1.5 = 12\,cm$$

11 $u = 4.0\,cm, f = 7.0\,cm, v = ?$

$$\frac{1}{u} + \frac{1}{v} = \frac{1}{f} \quad \text{so} \quad \frac{1}{v} = \frac{1}{u} - \frac{1}{f} = \frac{u-f}{fu}$$

$$v = \frac{fu}{u-f} = \frac{7.0 \times 4.0}{4.0 - 7.0} = -9.3\,cm$$

This shows the image is 9.3 cm on the same side as the object; it is upright and the rays of light only appear to cross, making it virtual.

12 $M = -\frac{v}{u} = \frac{-9.3}{4.0} = 2.3$

13 $M = \frac{h_i}{h_o}$

$$h_i = Mh_o = 2.3 \times 5.0 = 11.5\,cm$$

WS 10.2 PAGE 148

1 If $n_1 > n_2$ then the light will refract away from the normal; if $n_1 < n_2$ then the light will refract towards the normal.

2 a

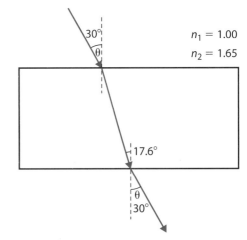

30°
θ
$n_1 = 1.00$
$n_2 = 1.65$
17.6°
θ
30°

$$n_1 \sin\theta_1 = n_2 \sin\theta_2$$

$$\sin\theta_2 = \frac{n_1 \sin\theta_1}{n_2}$$

$$\theta_2 = \sin^{-1}\left(\frac{n_1 \sin\theta_1}{n_2}\right) = \sin^{-1}\left(\frac{1.00 \sin 30°}{1.65}\right) = 17.6°$$

b **i** $n_1 = 1.00, \theta_1 = ?, \theta_2 = 18°$

Refractive index of the medium:

$$n_2 = \frac{c}{v_2} \therefore n_2 = \frac{3 \times 10^8}{2 \times 10^8} = 1.5$$

ii

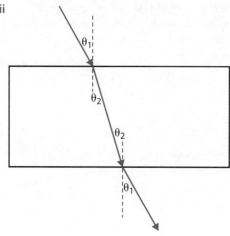

From part **i**, $n_2 = 1.50$

$$n_1 \sin\theta_1 = n_2 \sin\theta_2 \quad \text{so} \quad \sin\theta_1 = \frac{n_2 \sin\theta_2}{n_1}$$

$$\theta_1 = \sin^{-1}\left(\frac{n_2 \sin\theta_2}{n_1}\right) = \sin^{-1}\left(\frac{1.50 \sin 18°}{1.00}\right) = 27.6°$$

c $n_1 \sin\theta_1 = n_2 \sin\theta_2 \quad \text{so} \quad \sin\theta_1 = \frac{n_2 \sin\theta_2}{n_1}$

$\sin 90° = 1$, so

$$\sin\theta_c = \frac{n_2}{n_1} = \frac{1.33}{1.50}$$

$$\theta_c = \sin^{-1}\left(\frac{1.33}{1.50}\right) = 63°$$

3 By sending pulses of light down the cable it can transfer information accurately at close to the speed of light over greater distances than copper wire. It also allows for multiple beams of light to be sent down the cable, allowing for greater bandwidth.

4 By using total internal reflection. The core of the cable is made of glass and it is insulated by a material with a lower refractive index than the core. This allows for total internal reflection along the length of the fibre, preventing any of the light containing information from leaving the core.

WS 10.3 PAGE 150

1 The inverse square law shows the even spread of energy from a point source in three-dimensional space. As this will be moving in an expanding sphere away from the point source and the surface of a sphere is proportional to the radius squared ($4\pi r^2$), we can see that at any point facing the source its intensity will be inversely proportional to the square of distance from the source. This can also be shown by the following mathematical relationships:

2 The amount of graph paper illuminated, and the intensity of the light, will obey the inverse square law.

3 a

Distance from bulb (cm)	Number of squares illuminated	Area illuminated (cm^2)	Relative brightness (cm^{-2})
10	4	1.00	1.00
14	8.4	2.10	0.48
15	9.3	2.33	0.43
18	13.3	3.33	0.30
20	16.4	4.10	0.24
24	23.5	5.88	0.17
25	26	6.50	0.15
28	34.8	8.70	0.12
30	36.6	9.15	0.11

b

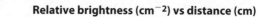
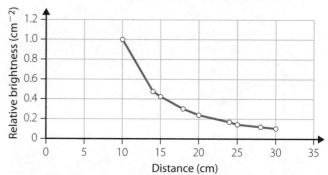

Relative brightness (cm^{-2}) vs distance (cm)

4 Yes, they show that the apparent brightness of the light follows an inverse square law relationship with distance from the light source.

5 This was to maintain the distance from the light source to the hole. This enabled that distance to be controlled and meant that only the independent variable of the distance from the light bulb was counted, making it a valid experiment.

Chapter 11: Thermodynamics

WS 11.1 PAGE 153

1 Temperature is a measure of the average kinetic energy of the particles in a substance.

2 The particles or atoms within an object will always have some vibrational energy regardless of their temperature. This vibrational energy increases as temperature increases, meaning that the particles will vibrate with a higher frequency and greater amplitude.

3 The Celsius scale of temperature is based around the freezing and boiling points of water at atmospheric pressure. This means that 0°C is equal to −273.15 K, and this tells us that the scales are NOT equivalent. However, a change in temperature (ΔT) of 1°C is equivalent to a change of 1 K. This means that we can measure the change in temperature using the Celsius scale and substitute this value into an equation wherever a change in temperature is required.

4 Because temperature is related to the average kinetic energy of particles, the more particles in a system the more accurate the measurement of temperature will be. In locations with very few particles (such as the upper atmosphere), the kinetic energy of particles can vary significantly and when an average is taken over very few particles it is not a good representation of the system.

9780170449595

5 a Particles in liquids are less tightly bound to other particles than are those in a solid, and as such they have rotational and translational energies. Particles in gases are not bound to each other and as such these energies are a larger component of their kinetic energy.

b Particles in solids are tightly bound. They can vibrate, but they are not free to rotate or move around so have only vibrational energy.

6 Steel and air will not contain the same amount of heat energy even when they have the same volume and temperature. This is due to the difference in heat capacity (the amount of energy required to raise the temperature of 1 kg of a substance by 1 K).

WS 11.2 PAGE 155

1 Equilibrium is a state in which all forces of physics are balanced.

2 In thermodynamics, equilibrium occurs when a system has reached a common temperature so there is no net transfer of energy. This is known as thermal equilibrium.

3 $T_t = \dfrac{m_1 T_1 + m_2 T_2}{m_t} = \dfrac{(1\,\text{kg})(73°C) + (1\,\text{kg})(25°C)}{2\,\text{kg}} = 49°C$

4 $T_2 = \dfrac{m_t T_t - m_1 T_1}{m_2} = \dfrac{(25.4\,\text{kg})(53.2°C) - (15.4\,\text{kg})(84°C)}{10\,\text{kg}}$
$= 5.8°C$

5 $T_t m_t = m_1 T_1 + m_2 T_2$ and $m_t = m_1 + m_2$. We need to find m_2.

Substituting the given values, we get:
$(40°C)(7.5 + m_2) = (80°C)(7.5) + (25°C)m_2$
$300 + 40\,m_2 = 600 + 25\,m_2$
$m_2 = 20\,\text{kg}$

6 $T_2 = \dfrac{m_t T_t - m_1 T_1}{m_2} = \dfrac{2(37.8°C) - 1(22.45°C)}{1} = 53.2°C$

WS 11.3 PAGE 156

1 This is due to there being differing number of particles within a unit mass of the substance and the different number of rotational, translational and vibrational modes for the particles within the substances in that state.

2 The high heat capacity of water allows it to absorb large amounts of thermal energy with minimal change in temperature. This allows water to absorb more thermal energy than most substances before reaching the temperature of the object it is cooling.

3 The high heat capacity of water means that large amounts of thermal energy must be transferred by the mirrors for the water to boil. This means that this system will take a long time to warm up each day before producing any electricity.

4 The lower specific heat capacity of the copper means that its temperature will increase more than that of the water for the same amount of thermal energy absorbed.

5 $Q = mc\Delta T = (0.2\,\text{kg})(4.18 \times 10^3\,\text{J}\,\text{kg}^{-1}\,\text{K}^{-1})(7.5\,\text{K}) = 6 \times 10^3\,\text{J}$

6 $\Delta T = \dfrac{Q}{mc} = \dfrac{8.46 \times 10^4\,\text{J}}{(2\,\text{kg})(0.385 \times 10^3\,\text{J}\,\text{kg}^{-1}\,\text{K}^{-1})} = 109\,\text{K or }109°C$

$T_f = T_i + \Delta T = 20°C + 110°C = 130°C$

7 $m = \dfrac{Q}{c\Delta T} = \dfrac{4.72 \times 10^4\,\text{J}}{(4.18 \times 10^3\,\text{J}\,\text{kg}^{-1}\,\text{K}^{-1})(78\,\text{K})} = 0.14\,\text{kg}$

8 $Q_{\text{water}} = m_{\text{water}} c_{\text{water}} \Delta T_{\text{water}}$
$= (0.24\,\text{kg})(4.18 \times 10^3\,\text{J}\,\text{kg}^{-1}\,\text{K}^{-1})(14.3\,\text{K}) = 1.4 \times 10^4\,\text{J}$

The law of conservation of energy dictates that the thermal energy gained by the water is equal to the thermal energy lost by the metal. From here we can calculate the specific heat capacity of the metal.

$c_{\text{metal}} = \dfrac{Q_{\text{water}}}{m_{\text{metal}} \Delta T_{\text{metal}}} = \dfrac{1.4 \times 10^4\,\text{J}}{(0.735\,\text{kg})(160.2\,\text{K})}$
$= 1.2 \times 10^2\,\text{J}\,\text{kg}^{-1}\,\text{K}^{-1}$

9 The specific heat capacity of the oil is half that of water, meaning it would require half the energy to raise the temperature by 1°C. However, the temperature change the oil must undergo to boil is more than double that of water if we assume the initial temperature is 25°C ($\Delta T = 75°C$ for water and 253°C for the oil). This means that the total energy required to boil the same mass of oil to drive a turbine would be greater than that required to boil water.

WS 11.4 PAGE 158

1 When a substance undergoes a change of state the orientation and organisation of the particles change and the forces acting between these particles also change. Depending on the type of state change, these forces (or bonds) are either overcome (requiring energy input) or allowed to form (releasing energy).

2 Latent heat of change of state is the energy absorbed or released by a substance as it undergoes a phase change.

3 Both latent heat of fusion and latent heat of vaporisation refer to the energy required to change the state of a substance, but they refer to different phase changes. Fusion refers to the change between solid and liquid and vaporisation refers to the change between liquid and gas. Because they refer to different phase changes, they have different values. The energy required to turn solid water (ice) into liquid water at 0°C is *not* the same as the energy required to turn liquid water into gaseous water (steam) at 100°C. The latent heat of fusion of water is only $3.34 \times 10^3\,\text{J}\,\text{kg}^{-1}$ compared to $2.26 \times 10^6\,\text{J}\,\text{kg}^{-1}$ for latent heat of vaporisation.

6 Total heat $= (0.01\,\text{kg})(4.18 \times 10^3\,\text{J}\,\text{K}^{-1}\,\text{kg}^{-1})(10\,\text{K}) = 4.18 \times 10^2\,\text{J}$

7 Rate of transfer $= \dfrac{4.18 \times 10^3\,\text{J}}{\textit{Your measured time here in seconds}}$

WS 11.5 PAGE 162

1 Thermal conductivity is the rate at which thermal energy is transferred through a material.

2 $k = \dfrac{Qd}{tA\Delta T} = \dfrac{(\text{J})(\text{m})}{(\text{s})(\text{m}^2)(\text{K})} = \text{J}\,\text{s}^{-1}\,\text{m}^{-1}\,\text{K}^{-1}$, or joule per second per metre per kelvin.

3 Both thermal conductivity (k) and specific heat capacity (C) are intrinsic properties of a substance; however, they represent different properties. Specific heat capacity is a measure of how much energy can be stored in the vibrational modes of a substance, whereas thermal conductivity is a measure of how fast the vibrations pass energy to neighbouring particles within the length of a material. Both are constant for a constant temperature.

4 Thermal conductivity is a measure of the rate at which thermal energy is transferred from one place to another by means of conduction. The greater the cross-sectional area of the conduction path the more energy can be conducted simultaneously. Thus the cross-sectional area of the conduction path is proportional to the rate at which heat energy is conducted.

5 Materials with a high thermal conductivity could be used for heat sinks and to transport heat from one region to another. Most metals can be used as heat sinks, with diamond used for very high-end applications. Materials with a low thermal conductivity would be used as thermal insulators to separate hot and cold systems. These materials include aerated polystyrene, concrete and the noble gasses.

6 Thermal conductivity only relates to the conduction method of heat transfer. Fluids are capable of heat transfer by means of convection in addition to conduction, and it is difficult to identify how much heat has been transferred via each method.

7 We know that $\dfrac{Q}{t} = 120\,\text{J s}^{-1}$.

Rearranging $\dfrac{Q}{t} = \dfrac{kA\Delta T}{d}$ to find k we get $k = \dfrac{120d}{A\Delta T}$.

Substituting $\Delta T = 10\,\text{K}$, $d = 4 \times 10^{-3}\,\text{m}$ and

$A = 1.6 \times 10^{-3}\,\text{m}^2$, we get:

$$k = \dfrac{(120\,\text{J s}^{-1})(4 \times 10^{-3}\,\text{m})}{(1.6 \times 10^{-3}\,\text{m}^2)(10\,\text{K})}$$

$$= 30\,\text{J s}^{-1}\,\text{K}^{-1}\,\text{m}^{-1}$$

8 a First we need to look at how much heat is being absorbed by the air. Given we know its mass, specific heat capacity and the temperature difference this is straightforward.

$Q = mC\Delta T = (31.25\,\text{kg})(718\,\text{J kg}^{-1}\,\text{K}^{-1})(5\,\text{K})$

$= 1.1 \times 10^5\,\text{J}$

Remember that a change of 1°C is equal to a change of 1 K.

Now that we have the amount of heat being transferred through the insulative material we can use $\dfrac{Q}{t} = \dfrac{kA\Delta T}{d}$ to determine how long this will take. Rearranging gives us:

$t = \dfrac{Qd}{kA\Delta T} = \dfrac{(1.1 \times 10^5\,\text{J})(6.5 \times 10^{-2}\,\text{m})}{(5.65 \times 10^{-3}\,\text{J s}^{-1}\,\text{m}^{-1}\,\text{K}^{-1})(17\,\text{m}^2)(25\,\text{K})}$

$= 3.0 \times 10^3\,\text{s}$ (to 2 significant figures)

b $t = \dfrac{Q}{kA\Delta T}d$

$t \propto d$

If the thickness was tripled, the time taken to heat up would also triple.

MODULE THREE: CHECKING UNDERSTANDING PAGE 164

1 **D** A 150 MHz radio wave

2 **C** The wavelength of the light will change

3 **D** 22°

4 **A** The ray lines cross at a single point

5 **B** The sum of the contributing waves

6 a

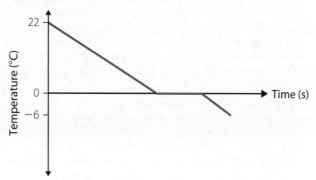

b Explanation should include the specific heat capacity of water and the latent heat capacity of water and ice.

7 $Q_{\text{in}} = Q_{\text{out}} \therefore m_w c_w \Delta T_w = -m_{Al} c_{Al} \Delta T_{Al}$

$\therefore m_w c_w (T_f - T_{i,w}) = m_{Al} c_{Al}(T_{i,Al} - T_f)$

$\therefore m_w c_w T_f - m_w c_w T_{i,w} = m_{Al} c_{Al} T_{i,Al} - m_{Al} c_{Al} T_f$

$\therefore T_f = \dfrac{m_{Al} c_{Al} T_{i,Al} + m_w c_w T_{i,w}}{m_w C_w + m_{Al} C_{Al}}$

$= \dfrac{20\,\text{kg} \times 900\,\text{J kg}^{-1}\,°\text{C}^{-1} \times 85°\text{C} + 50\,\text{kg} \times 4200\ \text{J kg}^{-1}\,°\text{C}^{-1} \times 15°\text{C}}{50\,\text{kg} \times 4200\,\text{J kg}^{-1}\,°\text{C}^{-1} + 20\,\text{kg} \times 900\,\text{J kg}^{-1}\,°\text{C}^{-1}}$

$T_f = 20.5°\text{C}$

MODULE FOUR: ELECTRICITY AND MAGNETISM

REVIEWING PRIOR KNOWLEDGE PAGE 166

1 The electron is responsible for the negative charge and the proton is responsible for the positive charge.

2 Like charges will experience a force of repulsion; unlike charges will experience a force of attraction.

3 Voltage is a measure of how much force is applied to charges in a circuit.

Current is a measure of how many charges pass a point per unit of time.

Resistance is a measure of how difficult it is for charges to move through a circuit.

4 Ohm's law states that the current flowing through a conductor is directly proportional to voltage applied to the conductor and can be written as $V = IR$ where R is the resistance of the conductor.

5 An electrical conductor is a material through which electrical charge can pass relatively easily. Metals such as copper and aluminium are examples of electrical conductors.

An electrical insulator is a material through which electrical charge cannot pass easily. Plastic and air are examples of electrical insulators.

6 In a series circuit all charge must pass through all components in sequence. A parallel circuit provides multiple paths through which the charge can move.

7 Iron is an example of a magnetic material.

8 a Two north poles repel because the field lines are in opposite directions and therefore compress.

b A north and south pole attract each other because the field lines point in the same direction and can link up.

9 A navigation compass works by using a small magnet that aligns itself with Earth's magnetic field. (The north pole of the compass is actually the south pole of the magnet.)

Chapter 12: Electrostatics

WS 12.1 PAGE 168

1 Matter consists of positive and negative charges in the form of protons and electrons, and uncharged neutrons. Objects become charged when there is an imbalance in these charged particles. This is achieved when one form of charge is removed from an object due to contact with another object or interaction with an electric field. Typically, it is the negative charge (electron) that is moved, making an object positive (removing electrons) or negative (adding electrons); however, protons can also be transferred to create an imbalance of charge.

2 The electrostatic field is field force like the gravitational force but much stronger. It is a force that affects charge (both positive and negative) and can be attractive or repulsive based on the type of charge. Two objects with charge of the same sign will experience a repulsive force, whereas two objects with charge of opposite signs will experience an attractive force.

3 The first conclusion is very accurate because it is consistent with theory. The relationship is a qualitative one, so qualitative results are very suitable.

9780170449595

The second conclusion is somewhat accurate as while theory indicates that as separation distance increases force will decrease, this is a quantitative relationship and so the qualitative results do not provide any insight to the degree of inverse proportionality. (Electrostatic force is actually inversely proportional to the square of the separation distance.)

The third conclusion is also somewhat accurate as once again the general relationship is consistent with theory. Quantitative data would be needed to confirm the linear proportionality between magnitude of charge and electrostatic force.

4 Separation distance:

1 Set up an electric field sensor near a charge of known value (e.g. one produced by a Van der Graaff generator).

2 Place the sensor at a distance of 1 m from the charged surface and record the value.

3 Repeat step 2, moving the sensor closer to the charged surface in 0.05 m increments.

Repeat steps 2 and 3 three more times to produce consistent trials for improved reliability.

WS 12.2 PAGE 170

1 The direction of an electric field is defined as the direction of the force vector that is applied to a positive point charge.

2 The relative spacing of the electric field lines represents the relative strength of the field. More closely spaced lines indicate a stronger electric field and more sparsely placed lines indicate a weaker electric field.

3 If field lines were to cross it would indicate that there is a force acting in two directions simultaneously.

4 Lines of equipotential cut the electric field lines at right angles. Each equipotential line connects all points that have the same electric potential; this means that there is no potential difference between any points on a equipotential line and therefore no work is done in moving a charge along this line.

5

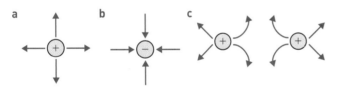
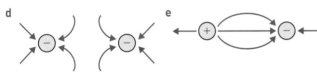
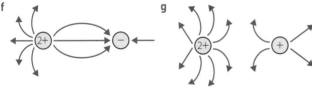
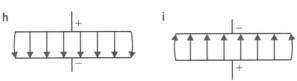

6 When the distance between point charges is doubled the force is reduced by a factor of 2^2. This is due to the radial nature of the field. If the distance between parallel plates is doubled this leads to a halving of the force, due to the linear nature of this field.

7 Parallel plates are not infinite, and the ends or edges of charged plates begin to resemble or approximate spherical charge. As such, the field lines would be radial (still perpendicular to the charge surface); this causes the outermost field lines of one plate to be directed outwards and then curve back around to meet the other plate.

WS 12.3 PAGE 172

1 $\vec{F} = \vec{E}q = (4.0 \times 10^3 \, \text{N C}^{-1})(-1.602 \times 10^{-19} \, \text{C}) = -6.4 \times 10^{-16} \, \text{N}$

$= 6.4 \times 10^{-16} \, \text{N}$ down the page

(Note the negative sign on the charge indicates that it would move opposite to the direction of the field.)

2 $\vec{E} = \dfrac{\vec{F}}{q} = \dfrac{6.42 \times 10^{-11} \, \text{N}}{1.602 \times 10^{-19} \, \text{C}} = 4.01 \times 10^{-8} \, \text{N C}^{-1}$ to the right

3 $\vec{F}_\alpha = \vec{E}q = (1.00 \times 10^5 \, \text{N C}^{-1})(3.204 \times 10^{-19} \, \text{C})$

$= 3.20 \times 10^{-14} \, \text{N}$ to the right

$\vec{F}_\beta = \vec{E}q = (1.00 \times 10^5 \, \text{N C}^{-1})(-1.602 \times 10^{-19} \, \text{C})$

$= -1.60 \times 10^{-14} \, \text{N} = 1.60 \times 10^{-14} \, \text{N}$ to the left

Note that the negative sign on the charge indicates that the electron would move in the opposite direction to the field.

4 $\vec{F} = \vec{E}q = (465 \, \text{N C}^{-1})(-1.602 \times 10^{-19} \, \text{C}) = -7.45 \times 10^{-17} \, \text{N}$

$= 7.45 \times 10^{-17} \, \text{N}$ towards the positive plate

$a = \dfrac{\vec{F}}{m} = \dfrac{7.45 \times 10^{-17} \, \text{N}}{9.109 \times 10^{-31} \, \text{kg}} = 8.18 \times 10^{13} \, \text{m s}^{-2}$ towards the positive plate

5 $\vec{F} = \dfrac{q_1 q_2}{4\pi\varepsilon_0} \dfrac{1}{r^2}$

If $\dfrac{q_1 q_2}{4\pi\varepsilon_0}$ is constant then $\vec{F} \propto \dfrac{1}{r^2}$

$\therefore 16\vec{F} \propto \dfrac{1}{\left(\dfrac{r}{4}\right)^2}$

If the distance is reduced by a factor of 4 then the force will be increased by a factor of 16, so the charges each experience a force of 16F.

6 $q^2 = r^2 F 4\pi\varepsilon_0$

If $F4\pi\varepsilon_0$ is constant, then $q^2 \propto r^2$.

Therefore, if the charge is doubled the distance that would provide the same force would also be doubled, so $2d$ would provide the same force.

7 $24 \, \text{mm} = 2.4 \times 10^{-2} \, \text{m}$ and $3.942 \, \text{nC} = 3.942 \times 10^{-9} \, \text{C}$

$\vec{F} = \dfrac{1}{4\pi\varepsilon_0} \dfrac{q_1 q_2}{r^2}$

$= \dfrac{1}{1.1126 \times 10^{-10} \, \text{C}^2 \, \text{N}^{-1} \, \text{m}^{-2}} \dfrac{(3.942 \times 10^{-9} \, \text{C})(-1.602 \times 10^{-19} \, \text{C})}{(2.4 \times 10^{-2} \, \text{m})^2}$

$= -9.9 \times 10^{-5} \, \text{N}$

$= 9.9 \times 10^{-5} \, \text{N}$ towards the point-like charge

8 $12 \, \text{cm} = 1.2 \times 10^{-1} \, \text{m}$

Combining $\vec{F} = m\vec{a}$ and $\vec{F} = \dfrac{1}{4\pi\varepsilon_0} \dfrac{q_1 q_2}{r^2}$, we get

$m\vec{a} = \dfrac{1}{4\pi\varepsilon_0} \dfrac{q_1 q_2}{r^2}$.

Rearranging gives $q_1 = \dfrac{4\pi\varepsilon_0 r^2 m\vec{a}}{q_2}$

$q_1 = \dfrac{(1.1126 \times 10^{-10} \, \text{C}^2 \, \text{N}^{-1} \, \text{m}^{-2})(2.1 \times 10^{-6} \, \text{m})^2 (1.673 \times 10^{-27} \, \text{kg})(9.8 \, \text{N kg}^{-1})}{1.602 \times 10^{-19} \, \text{C}}$

$= +1.6 \times 10^{-19} \, \text{C}$

9 $7.4 \, \mu\text{N} = 7.4 \times 10^{-6} \, \text{N}$

$r = \sqrt{\dfrac{1}{4\pi\varepsilon_0} \dfrac{q_1 q_2}{\vec{F}}} = \sqrt{\dfrac{1}{1.1126 \times 10^{-10} \, \text{C}^2 \, \text{N}^{-1} \, \text{m}^{-2}} \dfrac{1.5 \times 10^{-72} \, \text{C}^2}{7.4 \times 10^{-6} \, \text{N}}}$

$= 5.2 \, \text{m}$ (to 2 significant figures)

10 $d = \dfrac{V}{E} = \dfrac{12\,V}{400\,V\,m^{-1}} = 0.030\,m$

Note that in order to get the separation in metres we must use electric field strength units of $V\,m^{-1}$.

11 $5\,mm = 5 \times 10^{-3}\,m$

$V = Ed = (1.2 \times 10^3\,V\,m^{-1})(5 \times 10^{-3}\,m) = 6\,V$

12 $20\,\mu m = 2.0 \times 10^{-5}\,m$

$E = \dfrac{V}{d} = \dfrac{5\,V}{2.0 \times 10^{-5}\,m} = 2.5 \times 10^5\,V\,m^{-1}$

WS 12.4 PAGE 175

1 $W = Fd$ (1) and $\vec{F} = \vec{E}q$ (2)

Substituting equation (2) into (1) we get:

$W = Eqd$

2 $Eqd = -\Delta U$ (1) and $E = \dfrac{V}{d}$ (2)

Substituting equation (2) into (1) we get $\dfrac{V}{d}qd = -\Delta U$

So $V = \dfrac{-\Delta U}{q}$

3 Lines of equipotential cut the electric field lines at right angles. Each equipotential line connects all points that have the same electric potential.

4 If there is no change in potential ($\Delta V = 0$) then there must be no change in potential (or kinetic) energy of a charge, regardless of how far the charge moves.

5 $5.0\,mm = 5.0 \times 10^{-2}\,m$

$\Delta U = Eqd = (2.54 \times 10^{-6}\,N\,C^{-1})(-1.602 \times 10^{-19}\,C)(5.0 \times 10^{-2}\,m)$
$= -2.0 \times 10^{-26}\,N\,m = -2.0 \times 10^{-26}\,J$

Note that $N\,m$ when used in the context of work is equal to the joule (J).

6 $\Delta V = \dfrac{\Delta U}{q} = \dfrac{8.19 \times 10^{-16}\,J}{1.602 \times 10^{-19}\,C} = 5.11 \times 10^3\,J\,C^{-1}$

$= 5.11 \times 10^3\,V$

Note that a volt is equal to a joule per coulomb.

7 An electron volt is a measure of the kinetic energy gained or lost when an electron is accelerated across a potential difference of $1\,V$.

If we take the fundamental charge to be e, then substituting in $\Delta U = Vq$ gives $\Delta U = (1\,V)(1\,e) = 1\,eV$.

So electron volt is a measure of energy and can be used to simplify the answer where the change in energy is exceptionally small.

$\Delta U = (1\,V)(-1.602 \times 10^{-19}\,C) = -1.602 \times 10^{-19}\,V = -1.602 \times 10^{-19}\,J$

So $1\,eV = 1.602 \times 10^{-19}\,J$

8 $\Delta E_k = \dfrac{1}{2}mv^2$ and $Eqd = \Delta E_k \therefore Eqd = \dfrac{1}{2}mv^2$

Rearranging gives us

$d = \dfrac{mv^2}{2Eq} = \dfrac{(1.672 \times 10^{-27}\,kg)(2.5 \times 10^6\,m^2\,s^{-2})}{2(1.0 \times 10^4\,N\,C^{-1})(1.602 \times 10^{-19}\,C)}$

$= 3.3\,\dfrac{kg\,m^2\,s^{-2}}{N\,C^{-1}\,C}$

Looking at just the units, we have $\dfrac{kg\,m^2\,s^{-2}}{N\,C^{-1}\,C}$. The C and C^{-1} cancel, and from $F = ma$ we can substitute $kg\,m\,s^{-2}$ for N. This leaves us with $\dfrac{kg\,m^2\,s^{-2}}{kg\,m\,s^{-2}}$, and cancelling out the $kg\,m\,s^{-2}$ leaves us with units of metres. So $d = 3.3\,m$.

9 $mg = \dfrac{qV}{d}$

Rearranging gives $V = \dfrac{mgd}{q} = \dfrac{(1.673 \times 10^{-27})(9.8)(0.001)}{1.602 \times 10^{-19}}$

$= 1.0234 \times 10^{-10}\,V$

Chapter 13: Electric circuits

WS 13.1 PAGE 177

1 Conductors are materials through which a current can pass easily. They have low internal resistance, which is typically due to 'free' electrons within their structure.

Insulators are materials that have very high internal resistance. This resistance is typically due to a lack of free electrons in their structure, meaning the electrons are tightly bound. Some insulators will pass a current for extremely high voltages, but their structure often breaks down before this can occur. The resistance of a material will often increase with its temperature.

2 A current is a flow of charged particles. It can be in the form of direct or alternating current and can include positive and negative charges.

3 Current is a measure of the amount of charge passing a point per second. It has units of amperes or amps (A) and is denoted with the symbol I. The unit for charge is the coulomb (C). The number of coulombs per second flowing in a circuit is defined as the current.

4 An electron has a charge of $1.602 \times 10^{-19}\,C$, so one coulomb per second is equivalent to a flow of $\dfrac{1}{1.602 \times 10^{-19}} = 6.242 \times 10^{18}$ electrons per second.

5 $q = It = (0.45\,A)(72\,s) = 32\,C$

6 $t = \dfrac{q}{I} = \dfrac{7.62 \times 10^6\,C}{1.60\,A} = 4.76 \times 10^6\,s$

7 time $= 42$ minutes $= (42 \times 60)\,s = 2520\,s$

$I = \dfrac{q}{t} = \dfrac{2.3 \times 10^3\,C}{2520\,s} = 0.91\,A$

8 a Current is a measure of the amount of charge that passes a point per unit of time or second, therefore current can be considered the rate of movement of charge.

   b For a constant voltage (or constant applied electric field), copper allows the most charge to flow through it per second. This is due to copper having more free electrons per unit volume than the other metals. This is followed by aluminium and magnesium, with iron allowing the smallest amount of charge to flow per second.

9 This is a common way for battery capacity to be noted and means that, at the prescribed voltage, the battery can deliver $5.2\,A$ of current continuously for 1 hour (or $1\,A$ for $5.2\,h$).

Given $1\,A$ is $1\,C$ per second, the total charge would be equal to $(5.2)(3600) = 18\,720\,C$.

WS 13.2 PAGE 179

1 $I = \dfrac{V}{R} = \dfrac{240\,V}{425\,\Omega} = 0.56\,A$

2 $V = IR = (1.00\,A)(1465\,\Omega) = 1465\,V = 1.47 \times 10^3\,V$

3 $R = \dfrac{V}{I} = \dfrac{110\,V}{7.2\,A} = 15.28\,\Omega = 15\,\Omega$

4 $R = \dfrac{V}{I} = \dfrac{12\,V}{0.15\,A} = 8.0\,\Omega$

5 a $R_{old} = \dfrac{V}{I} = \dfrac{240\,V}{9.6\,A} = 25\,\Omega$

      $R_{new} = (0.8)(25\,\Omega) = 20\,\Omega$

9780170449595

b $R_{old} - R_{new} = 25\,\Omega - 20\,\Omega = 5\,\Omega$

6 $W = Vq = (240\,\text{V})(1.602 \times 10^{-19}\,\text{C}) = 3.8 \times 10^{-17}\,\text{J}$

$1\,\text{eV} = 1.602 \times 10^{-19}\,\text{J}$

$\therefore \dfrac{3.8 \times 10^{-17}\,\text{J}}{1.602 \times 10^{-19}\,\text{J}\,\text{eV}^{-1}} = 240\,\text{eV}$

7 $V = \dfrac{W}{q} = \dfrac{24.6\,\text{eV}}{1\text{e}} = 24.6\,\text{V}$

Note that we can use either joules and coulombs *or* electron volts and fundamental charges to calculate potential difference.

8 $W = Vq = (18\,\text{V})(2.4\,\text{C}) = 43\,\text{J}$

9 $q = \dfrac{W}{V} = \dfrac{725\,\text{J}}{24\,\text{V}} = 30\,\text{C}$

10 $V = \dfrac{W}{q} = \dfrac{74.52\,\text{J}}{1.84 \times 10^{-2}\,\text{C}} = 4050\,\text{V}$

11 a

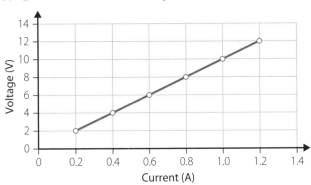

Circuitry resistor

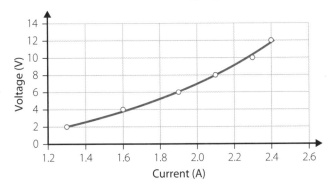

Incandescent light globe

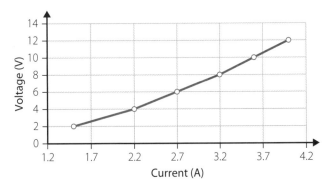

Nichrome wire

b Circuitry resistor – ohmic; incandescent light globe – non ohmic; nichrome wire – non-ohmic.

c The resistance of the non-ohmic components increased due to a factor independent of voltage or current, such as temperature. As we learnt in chapter 11, an increase in temperature indicates greater vibrational energy within the metal lattice; this may impede the flow of electrons further.

12 a Exponential

b The current will drop to zero.

c Ohm's law can be applied to model resistance in basic conductors and resistors used within a circuit and allows for quick and simple calculations to be made. Its limitations are found when non-ohmic components such as diodes are considered. Given the wide applications of non-ohmic components in modern circuitry, the usefulness of Ohm's law in the understanding of more complex circuits is limited.

WS 13.3 PAGE 182

1 $P = \dfrac{W}{t}$ Substituting $W = Vq$, we get $P = \dfrac{Vq}{t}$ but $\dfrac{q}{t} = I$;

therefore $P = VI$.

2 $q = \dfrac{W}{V} = \dfrac{145\,\text{J}}{(1\,\text{s})(4.2\,\text{V})} = 35\,\text{C}$

3 $E_{tot} = Pt = (65\,\text{W})(2.4 \times 60 \times 60\,\text{s}) = 5.6 \times 10^{5}\,\text{J}$

Since the globe converts 88% of the electrical energy to heat, $E_{heat} = (5.6 \times 10^{5}\,\text{J})(0.88) = 4.9 \times 10^{5}\,\text{J}$

4 $Q = mC\Delta T = (0.875\,\text{kg})(4.18 \times 10^{3}\,\text{J}\,\text{K}^{-1}\,\text{kg}^{-1})(73\,\text{K}) = 2.7 \times 10^{5}\,\text{J}$

$P = \dfrac{E}{t} = \dfrac{2.7 \times 10^{5}\,\text{J}}{(2.33 \times 60\,\text{s})} = 1.9 \times 10^{3}\,\text{W} = 1.9\,\text{kW}$

5 $I = \dfrac{P}{V} = \dfrac{1.9 \times 10^{3}\,\text{W}}{240\,\text{V}} = 7.9\,\text{A}$. The kettle will not overload the circuit.

6 a 1 Connect the coil of nichrome wire in series with a variable power source, an ammeter and a variable resistor.

2 Connect a voltmeter in parallel across the nichrome wire coil.

3 Add 200 g of water (measured with an electronic balance) to a polystyrene cup.

4 Fully submerge the coil in the water.

5 Measure the initial temperature of the water with a thermometer and record in a table.

6 Set the variable power supply to 24 V.

7 Switch on the variable power supply and start a stopwatch simultaneously.

8 Measure and record the temperature of the water every minute for 8 consecutive minutes.

b $E = Pt = VIt = (24\,\text{V})(5.2\,\text{A})(480\,\text{s}) = 60\,000\,\text{J} = 60\,\text{kJ}$

c $Q = mC\Delta T = (0.200\,\text{kg})(4.18 \times 10^{3}\,\text{J}\,\text{kg}^{-1}\,\text{K}^{-1})(66\,\text{K})$
$= 55\,000\,\text{J} = 55\,\text{kJ}$

d Energy cannot be created or destroyed, therefore the 5000 J difference would indicate that energy was lost to the environment. As we learnt in chapter 11, heat will flow between objects at different temperature to achieve thermal equilibrium. This is not only the case with the nichrome wire and the water, but also with the water, polystyrene cup and the surrounding atmosphere. This means that a portion of the heat energy is lost to the surrounding environment. There may also be energy losses within the other components of the circuit.

e They could repeat their method, keeping all variables controlled to obtain consistent data.

f An LED converts electrical energy into light energy.

Loudspeakers convert electrical energy into sound energy (a form of kinetic energy).

A motor converts electrical energy into rotational mechanical energy (a form of kinetic energy).

7 Energy efficiency is a measure of how much of the initial energy is converted into a useful or desired form of energy. It is often quoted as a percentage calculated by dividing the useful output energy by the total input energy.

8 The law of conservation of energy dictates that energy cannot be created or destroyed. This implies that when one form of energy is consumed, such as chemical or electrical potential energy, the sum of the energies produced must be equal. For the internal combustion engine vehicle, this means that while 35% of the chemical potential energy is converted into rotational kinetic energy, the remaining 65% of the chemical potential energy consumed must have also been converted into some other form. This 65% however was converted into undesirable forms of energy such as heat, light and other kinetic energies such as sound and vibrations.

The same laws apply to the electric vehicle; however, a much smaller portion (10%) is converted into undesirable forms of energy.

9 An air conditioner works by moving heat energy from one place to another. This is not the same as a standard heater, which converts electrical energy into heat energy. By using an air conditioner the 5.54×10^6 J of heat energy has simply been moved to another location rather than converted. The 2.75×10^4 J is the work done to move the heat energy. In this scenario, energy has not been created or destroyed and therefore it does not violate conservation laws.

WS 13.4 PAGE 185

1 $V_{tot} = V_1 + V_2 + V_3 + ... + V_n$

Ohm's law gives $V = IR$

Substituting, $I_{tot}R_{tot} = I_1R_1 + I_2R_2 + ... + I_nR_n$

We know that the total net movement of charge in a circuit is 0. Also, in a series circuit all the current must pass through each resistor. Given all charge must pass through each resistor, we know that $I_{tot} = I_1 = I_2 = ... = I_n$.

Dividing our equation by I_{tot} we get

$R_{tot} = R_1 + R_2 + ... + R_n$

2 $I_{tot} = I_1 = I_2 = ... = I_n$. From Ohm's law, $I = \dfrac{V}{R}$ and substituting

gives $\dfrac{V_{tot}}{R_{tot}} = \dfrac{V_1}{R_1} + \dfrac{V_2}{R_2} + \dfrac{V_3}{R_3} + ... + \dfrac{V_n}{R_n}$.

Given each unit of charge must transfer the same amount of energy regardless of the path, we know that the potential difference across each resistor is the same as the potential difference across the battery.

$V_{tot} = V_1 = V_2 = V_3 = ... = V_n$

From this we can divide both sides by V_{tot}, leaving us with the following expression:

$\dfrac{1}{R_{tot}} = \dfrac{1}{R_1} + \dfrac{1}{R_2} + \dfrac{1}{R_3} + ... + \dfrac{1}{R_n}$

3 As the resistors are in parallel, we can use:

$\dfrac{1}{R_{tot}} = \dfrac{1}{R_1} + \dfrac{1}{R_2} + \dfrac{1}{R_3} = \dfrac{1}{4.25\,\Omega} + \dfrac{1}{17.2\,\Omega} + \dfrac{1}{12.7\,\Omega}$

$= 0.372\,\Omega^{-1}$

$\therefore R_{tot} = \left(\dfrac{1}{0.372}\right)\Omega = 2.69\,\Omega$ (to 3 significant figures)

4 Here we can see that some resistors are in series with each other and some are in parallel. We need to consider these relationships separately.

R_1 and R_2 are in series; therefore:

$R_{1+2} = R_1 + R_2 = 0.25\,\Omega + 1.67\,\Omega = 1.92\,\Omega$

R_{1+2} and R_3 are in parallel, so we use:

$\dfrac{1}{R_{tot}} = \dfrac{1}{R_{1+2}} + \dfrac{1}{R_3} = \dfrac{1}{1.92\,\Omega} + \dfrac{1}{7.42\,\Omega} = 0.656\,\Omega^{-1}$

$\therefore R_{tot} = \dfrac{1}{0.656\,\Omega} = 1.52\,\Omega$ (to 3 significant figures)

5 First, we consider resistors in parallel, in this example R_2 and R_3. We can use

$\dfrac{1}{R_{2+3}} = \dfrac{1}{R_2} + \dfrac{1}{R_3} = \dfrac{1}{7.9\,\Omega} + \dfrac{1}{14.2\,\Omega} = 0.197\,\Omega^{-1}$

$\therefore R_{2+3} = \left(\dfrac{1}{0.197}\right)\Omega = 5.076\,\Omega$

Now, R_1 and R_{2+3} are in series, so we can use:

$R_1 = R_{tot} - R_{2+3} = 16.7\,\Omega - 5.076\,\Omega = 11.624\,\Omega = 11.6\,\Omega$

6 To find the current through the circuit we must first find the total equivalent resistance.

First, consider R_2 and R_3, which are in parallel.

$\dfrac{1}{R_{2+3}} = \dfrac{1}{R_2} + \dfrac{1}{R_3} = \dfrac{1}{13.50\,\Omega} + \dfrac{1}{6.47\,\Omega} = 0.229\,\Omega^{-1}$

$\therefore R_{2+3} = \left(\dfrac{1}{0.229}\right)\Omega = 4.37\,\Omega$

Now, R_1 and R_{2+3} are in series, so

$R_{1+(2+3)} = R_1 + R_{2+3} = 2.75\,\Omega + 4.37\,\Omega = 7.12\,\Omega$

R_5 and R_6 are in parallel:

$\dfrac{1}{R_{5+6}} = \dfrac{1}{R_5} + \dfrac{1}{R_6} = \dfrac{1}{20\,\Omega} + \dfrac{1}{25.1\,\Omega} = 0.090\,\Omega^{-1}$

$\therefore R_{5+6} = \left(\dfrac{1}{0.090}\right)\Omega = 11\,\Omega$

Now, R_4 and R_{5+6} are in series, so

$R_{4+(5+6)} = R_4 + R_{5+6} = 16.1\,\Omega + 11\,\Omega = 27\,\Omega$

$R_{1+(2+3)}$ and $R_{4+(5+6)}$ are in parallel, so

$\dfrac{1}{R_{tot}} = \dfrac{1}{R_{1+(2+3)}} + \dfrac{1}{R_{4+(5+6)}} = \dfrac{1}{7.12\,\Omega} + \dfrac{1}{27\,\Omega} = 0.18\,\Omega^{-1}$

$\therefore R_{tot} = \left(\dfrac{1}{0.18}\right)\Omega = 5.6\,\Omega$

Finally, $I = \dfrac{V}{R}$ gives us $\dfrac{12\,V}{5.6\,\Omega} = 2.1\,A$

7 First, consider R_3 and R_4, which are in parallel.

$\dfrac{1}{R_{3+4}} = \dfrac{1}{R_3} + \dfrac{1}{R_4} = \dfrac{1}{7.41\,\Omega} + \dfrac{1}{2.81\,\Omega} = 0.491\,\Omega^{-1}$

$\therefore R_{3+4} = \left(\dfrac{1}{0.491}\right)\Omega = 2.04\,\Omega$

Now R_{3+4} and R_{1+2} are in series, so

$R_{1+2} = R_{tot} - R_{3+4} = 5.2\,\Omega - 2.04\,\Omega = 3.16\,\Omega$

R_1 and R_2 are in parallel:

$\dfrac{1}{R_1} = \dfrac{1}{R_{1+2}} - \dfrac{1}{R_2} = \dfrac{1}{3.16\,\Omega} - \dfrac{1}{3.84\,\Omega} = 0.056\,\Omega^{-1}$

$\therefore R_1 = \left(\dfrac{1}{0.056}\right)\Omega = 17.8\,\Omega = 18\,\Omega$

8 First, consider R_1 and R_2, which are in parallel.

$\dfrac{1}{R_{1+2}} = \dfrac{1}{R_1} + \dfrac{1}{R_2} = \dfrac{1}{5.718\,\Omega} + \dfrac{1}{9.12\,\Omega} = 0.285\,\Omega^{-1}$

$\therefore R_{1+2} = \left(\dfrac{1}{0.285}\right)\Omega = 3.15\,\Omega$

R_4 and R_5 are also in parallel:

$\dfrac{1}{R_{4+5}} = \dfrac{1}{R_4} + \dfrac{1}{R_5} = \dfrac{1}{7.865\,\Omega} + \dfrac{1}{12\,\Omega} = 0.21\,\Omega^{-1}$

9780170449595

$$\therefore R_{4+5} = \left(\frac{1}{0.21}\right)\Omega = 4.8\,\Omega$$

R_3 is in series with R_{1+2} and R_{4+5}, so

$R_3 = R_{tot} - R_{1+2} - R_{4+5} = 47\,\Omega - 3.51\,\Omega - 4.8\,\Omega = 38.7\,\Omega = 39\,\Omega$

Chapter 14: Magnetism

WS 14.1 PAGE 189

1. Ferromagnetic materials have large domains that are easily aligned when subject to an external magnetic field. As these domains begin to align, they produce a field parallel to the external field. This induces a strong force of attraction between the ferromagnetic material and the external magnetic field. Ferromagnetic materials will keep their domain alignments to some extent after being removed from a magnetic field. All permanent magnets are made from ferromagnetic material.

2. Ferromagnetic materials and non-ferromagnetic materials will be distinguished by this process as ferromagnetic materials will be attracted to the external magnetic field. The non-ferromagnetic materials will show no observable interactions.

3. Separation distance between the permanent magnet and the test material

   Cross-sectional area of each material

   Mass of each material

4. Magnetism is an intrinsic property of electrons, which are a part of all matter. The orientation (or spin) of each electron within an atom or structure provides a very small local magnetic field.

5. Ferromagnetic materials have a large proportion of unpaired electrons in their structure. In a bulk material these unpaired electrons can align, forming domains. Domains are regions in which there is a net magnetic moment. Across the entirety of a ferromagnetic material, the domains will often cancel each other out, leaving a ferromagnetic material that is not in itself a magnet.

6. Iron, nickel and cobalt are all ferromagnetic elements.

7.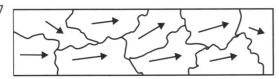

   Regions within the crystal structure have unpaired electrons that are aligned, producing a net magnetic moment. These regions (called domains) are aligned within a permanent magnet, increasing the net magnetic moment throughout the entire magnet. The more aligned electrons and domains, the stronger the magnetic field produced.

WS 14.2 PAGE 191

1. The direction of a magnetic field is defined as the direction in which the north end of a compass would move if placed within the field.

2. The relative spacing between field lines indicates relative field strength. The closer the field lines the stronger the field.

3. Field lines cannot cross as this would imply that there are two distinct values for the magnetic field at the same location and therefore a force acting in two separate directions simultaneously.

4.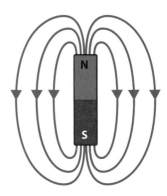

5. It would move in an antiparallel direction to the field lines.

6. a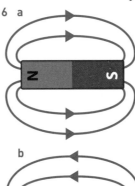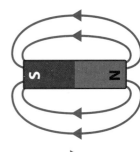

   b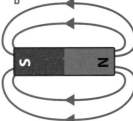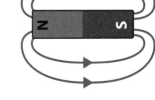

7. It is a point of zero field and a region in which either end of a compass needle would experience zero net force. This point of zero field is sometimes referred to as the null point.

8.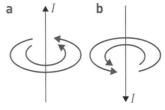

9.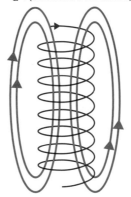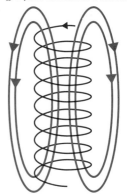

WS 14.3 PAGE 193

1 **1** Set up a series circuit with a variable power supply, conductor, variable resistor and ammeter.

  2 Set the power supply to 12 V and the variable resistor to its highest setting.

  3 Slowly decrease the resistance of the variable resistor until there is 1 A flowing through the circuit.

  4 Measure the magnetic field strength with a magnetic field probe at a distance of 0.10 m from the conductor.

  5 Repeat steps 4 and 5, increasing the current by 1 A each time to a maximum of 5 A. (Be sure to turn the circuit off in between measurements to avoid overheating.)

2

What are the risks in doing this investigation?	How can you manage these risks to stay safe?
Electric shock	Ensure all leads are well insulated and power supply is turned off when altering the components.
Burns from overheating wires	Ensure the circuit is only on while taking measurements or adjusting current values.

3

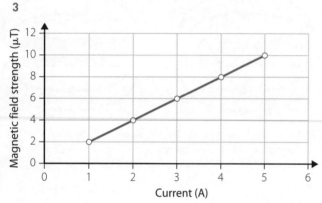

4 $B = \dfrac{\mu_0}{2\pi r} I$

If the gradient $= 2.0 \times 10^{-6}\,\text{T A}^{-1}$ then $2.0 \times 10^{-6} = \dfrac{\mu_0}{2\pi r}$.

$\mu_0 = (2.0 \times 10^{-6}\,\text{T A}^{-1})(2)(\pi)(0.10\,\text{m}) = 1.3 \times 10^{-6}\,\text{T m A}^{-1}$

(Note this is often written with units of H m, or Henry metre, or as N A^{-2}.)

5 The accuracy of the investigation was very high as the results obtained matched the expected or correct results.

6 If the value calculated above is from consistent (reliable) data, it would indicate a systematic error. This may have been due to improper calibration of the magnetic field strength probe (giving a consistently high reading) or consistently measuring the magnetic field strength of the conductor at a distance of less than 0.10 m.

7 gradient $= \dfrac{\text{rise}}{\text{run}} = \dfrac{\text{magnetic field strength in tesla}}{\text{current in amperes}}$

Substituting units, gradient $= \dfrac{\text{T}}{\text{A}}$

We know that T has base units of $\text{kg s}^{-2}\text{A}^{-1}$

Therefore gradient has units of $\text{kg s}^{-2}\text{A}^{-2}$

Now, $1\,\text{kg m s}^{-2} = 1\,\text{N}$

Therefore gradient has units of $\text{N A}^{-2}\text{m}^{-1}$

$1\,\text{H} = 1\,\text{N A}^{-2}$

Therefore gradient has units of H m^{-1}.

8 The relationship between field strength and distance is one that should be inversely proportional. This is a very difficult relationship to graph with limited data points. By graphing the

relationship between field strength and inverse distance a much simpler linear line of best fit can be applied.

9 Given a magnetic field is produced radially, the 2π is a scaling factor. The constant μ quantifies the degree to which a magnetic field can penetrate a given medium; μ_0 is the constant for free space and can be used as an approximation for air.

WS 14.4 PAGE 196

1 Magnetic field strength will increase linearly with increased number of coils.

2

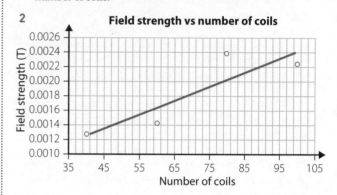

3 Gradient $= \dfrac{\text{Rise}}{\text{Run}} = \dfrac{0.0022\,\text{T} - 0.0016\,\text{T}}{89\,\text{coils} - 58\,\text{coils}} = 1.9 \times 10^{-5}\,\text{T coil}^{-1}$

4 We know that $\mu_0 = \dfrac{B}{N}\dfrac{L}{I}$.

Given the gradient of our graph above is $\dfrac{B}{N} = 1.9 \times 10^{-5}\,\text{T coil}^{-1}$, we can write an expression for μ_0 as

$\mu_0 = \text{gradient} \times \dfrac{L}{I} = (1.9 \times 10^{-5}\,\text{T coil}^{-1}) \times \dfrac{0.050\,\text{m}}{1.0\,\text{A}}$

$= 9.5 \times 10^{-7}\,\text{T m A}^{-1}$.

5 The experimental value is smaller than the accepted value. This is likely to be due to errors in the measurement of the magnetic field strength within the solenoid.

6 Coils may not have been uniform in circumference or in separation distance. This would result in a non-uniform magnetic field. The temperature of the coil (or variable resistor) may have increased over time, increasing resistance and reducing the current flowing and therefore reducing the magnetic field strength. The positioning of the sensor may not have been uniform across trials, introducing random error to the results.

WS 14.5 PAGE 198

1 $B = \dfrac{\mu_0 I}{2\pi r} = \dfrac{4\pi \times 10^{-7}\,\text{N A}^{-2}(2.4\,A)}{2\pi(0.25\,\text{m})} = 1.9 \times 10^{-7}\,\text{T}$

2 $r = \dfrac{\mu_0 I}{2\pi B} = \dfrac{4\pi \times 10^{-7}\,\text{N A}^{-2}(1.0\,A)}{2\pi(1.5 \times 10^{-6}\,\text{T})} = 0.32\,\text{m}$

3 $I = \dfrac{B 2\pi r}{\mu_0} = \dfrac{4.5 \times 10^{-6}\,\text{T} \times 2\pi(0.12\,\text{m})}{4\pi \times 10^{-7}\,\text{N A}^{-2}} = 2.7\,\text{A}$

4 $I = \dfrac{B 2\pi r}{\mu_0} = \dfrac{7.2 \times 10^{-5}\,\text{T} \times 2\pi(1.5\,\text{m})}{4\pi \times 10^{-7}\,\text{N A}^{-2}} = 540\,\text{A}$

5 $B = \dfrac{\mu_0 I}{2\pi r} = \dfrac{4\pi \times 10^{-7}\,\text{N A}^{-2}(15.4\,A)}{2\pi(0.45\,\text{m})} = 6.8 \times 10^{-6}\,\text{T}$

6 $I = \dfrac{B 2\pi r}{\mu_0} = \dfrac{2 \times 10^{-7}\,\text{T} \times 2\pi(0.05\,\text{m})}{4\pi \times 10^{-7}\,\text{N A}^{-2}} = 3 \times 10^{-2}\,\text{A}$

9780170449595

7 $N = \dfrac{Bl}{\mu_0 I} = \dfrac{(0.250\,\text{T})(0.080\,\text{m})}{4\pi \times 10^{-7}\,\text{N\,A}^{-2}(2.40\,\text{A})} = 6600$ turns

8 $B = \dfrac{\mu_0 NI}{l} = \dfrac{4\pi \times 10^{-7}\,\text{N\,A}^{-2}(270)(1.2\,\text{A})}{0.075\,\text{m}} = 5.4 \times 10^{-3}\,\text{T}$

$5.4 \times 10^{-3}\,\text{T} \times 1.2 = 6.5 \times 10^{-3}\,\text{T}$

$I = \dfrac{Bl}{N\mu_0} = \dfrac{(6.5 \times 10^{-3}\,\text{T})(0.075)}{4\pi \times 10^{-7}(270)} = 1.4\,\text{A}$

9 $25\ \text{turns cm}^{-1} = 2500\ \text{turns m}^{-1}$

$B = \dfrac{\mu_0 NI}{l} = \dfrac{4\pi \times 10^{-7}\,\text{N\,A}^{-2}(2500\,\text{turns m}^{-1})(0.2\,\text{A})}{1\,\text{m}}$

$\quad = 6 \times 10^{-4}\,\text{T}$

10 $\dfrac{B}{I\mu_0} = \dfrac{N}{l} = \dfrac{1.5 \times 10^{-3}}{(4\pi \times 10^{-7}\,\text{N\,A}^{-2})(1.75\,\text{A})} = 810\ \text{turns m}^{-1}$

$\quad = 8\ \text{turns cm}^{-1}$

Note that the number of turns must equal a whole number.

WS 14.6 PAGE 199

1 Graphical models allow students to understand the shape of magnetic fields. They also allow the relative field intensities to be demonstrated.

Mathematical models are beneficial as they provide approximate values for measurable variables and allow numerical predictions to be made prior to the construction of expensive equipment such as transformers. This in turn allows engineers to build equipment that will produce the correct magnetic fields required.

2 Graphical models do not provide any quantitative data about a given situation and are difficult to interpret for complex scenarios.

Mathematical models provide only approximate values for a given scenario and are limited in their accuracy by the precision of measurements made for each variable. With mathematical models, it can be difficult to visualise the relationship between variables.

3 A good model should be easy to interpret and provide information that reasonably approximates real-world scenarios.

4 Field forces (including gravitational, electrical and magnetic) and electromagnetic and mechanical waves

5 Models are an essential part of science in both education and research. While not perfect representations, they provide suitable data for the vast majority of situations and understanding of key principles involved.

MODULE FOUR: CHECKING UNDERSTANDING PAGE 200

1 **C** Diode

2 **A** $2.4\ \Omega$

3 **D** $F \propto \dfrac{1}{d^2}$

4 **A** The direction in which the south pole of a compass would point.

5 $F = \dfrac{1}{4\pi\varepsilon_0}\dfrac{q_1 q_2}{r^2}$

Substituting the charge on two helium nuclei we get

$F = \dfrac{(3.204 \times 10^{-19})(3.204 \times 10^{-19})}{(4)(\pi)(8.854 \times 10^{-12})(2.45 \times 10^{-12})^2} = 1.5371 \times 10^{-15}\,\text{N}$

$\quad = 1.54 \times 10^{-15}\,\text{N}$

6 a $V = \dfrac{\Delta U}{q}$

Rearranging we get $\Delta U = Vq$

Substituting we get

$\Delta U = (-12)(-1.602 \times 10^{-19}) = 1.9224 \times 10^{-18}\,\text{J} = 1.9 \times 10^{-18}\,\text{J}$.

Note: the question indicates that the field is accelerating the electron; therefore, the work done on the electron must be positive.

b $\Delta U = E_\text{k} = \dfrac{mv^2}{2}$

Rearranging we get $v = \sqrt{\dfrac{2E_\text{k}}{m}}$

Substituting we get

$v = \sqrt{\dfrac{(2)(1.9224 \times 10^{-18})}{9.109 \times 10^{-31}}} = 2.054\,478 \times 10^{8}\,\text{m\,s}^{-1}$

$\quad = 2.1 \times 10^{8}\,\text{m\,s}^{-1}$

7 $I = \dfrac{V}{R}$

Substituting we get $I = \dfrac{240}{27.5} = 8.727\,272\,\text{A}$

$P = VI$

Substituting we get $P = (240)(8.727\,272) = 2094.54\,\text{W}$

$\quad\quad\quad\quad\quad\quad\quad\quad\quad\quad\quad = 2100\,\text{W}$

8 First we need to convert 4.25 cm to 0.0425 m.

$F = \dfrac{\mu_0 NI}{L}$

Substituting we get $F = \dfrac{4NI}{L}$

$F = \dfrac{(4\pi \times 10^{-7})(275)(6.10)}{(0.0425)} = 0.049\,600\,2\,\text{N} = 4.96 \times 10^{-2}\,\text{N}$

9 The field intensity of an electromagnet can be varied by varying the current that passes through the coil, whereas the field intensity of permanent magnets is constant (although it can often reduce over time due to thermal energy rearranging domains). Electromagnets are less susceptible to damage by temperature or physical shock and therefore can be used in higher temperature situations and more robust applications.

10 Ferromagnetism arises from unpaired electron spins within a crystal structure. If a majority of unpaired spins in a region align, they produce domains within a crystal. If a majority of domains within the bulk material align, it will lead to ferromagnetism.